T0355992

Praise for Charan Rang...

WHY WE REMEMBER

"[Ranganath's] descriptions of complex studies are entertaining and clarifying, and he vividly paints the intellectual history of the science of memory. . . . He's a generous, humble narrator. . . . What's most compelling about *Why We Remember* is that it offers a scientifically robust rationale to accept with grace that, no matter what happens in this new world, we will not remember everything we want. Memory research makes clear that there is no use in fighting the tide of forgetting that leaves some memories ashore even as it sweeps away—mercifully, at times—the rest."

—*The Washington Post*

"It has never been easier to fact-check our memories against an external record and find ourselves lacking, but Ranganath is intent on giving us a new way of understanding memory."

—*The New Yorker*

"Ranganath is an astute and affable tour guide."

—*Undark* magazine

"A riveting overview of how memory works. . . . Ranganath has a knack for describing neuroanatomy in accessible terms, and the science consistently surprises. . . . Approachable and enlightening, this is worth seeking out."   —*Publishers Weekly* (starred review)

"Prominent neuroscientist and Guggenheim fellow Charan Ranganath guides us through the science of our memories with incredible insight and clear science. He combines fascinating tales of the peculiarities of memory with practical, actionable steps. Not only will every reader remember better afterward, they'll also never forget this life-changing book."   —Siddhartha Mukherjee,
Pulitzer Prize–winning author of
The Emperor of All Maladies and *The Gene*

"*Why We Remember* offers a radically new and engaging explanation of how and why we remember. More than just a record of our past, Dr. Ranganath shows us that memories are deeply involved in the present and a path toward an anticipated future. It is a tour de force of both individual and collective importance."

—Dr. Matthew Walker, author of *Why We Sleep*

"In this magnum opus, leading memory researcher Charan Ranganath turns much of what we think we know about memory on its head, revealing through hard evidence that the primary mission of our brain's memory system is, in many respects, to forget things in order to prepare us for a changing and uncertain future. Ranganath is a master explainer and storyteller."

—Daniel J. Levitin, author of
Successful Aging and *This Is Your Brain on Music*

"*Why We Remember* is terrific. Ranganath balances original first-class science with lighter, more personal writing. This will be a mind-changing read for anyone who wants to better understand and use their own brain." —Robert M. Sapolsky, author of *Behave*

"*Why We Remember* is going to transform readers' understanding of memory. It's hard to think of a topic more timely and important to communicate to the world or a scientist who is better positioned to do so than Charan Ranganath."

—Ethan Kross, author of *Chatter*

Charan Ranganath, PhD

WHY WE REMEMBER

Charan Ranganath is a professor at the Center for Neuro-science and Department of Psychology and the director of the Memory and Plasticity Program and the Dynamic Memory Lab at the University of California at Davis. For more than twenty-five years, Dr. Ranganath has studied the mechanisms in the brain that allow us to remember past events, using brain imaging techniques, computational modeling, and studies of patients with memory disorders. He has been recognized with a Guggenheim Fellowship, and his writing has appeared in *The New York Times*, *The Wall Street Journal*, the *Los Angeles Times*, and elsewhere. He lives in Davis, California.

charanranganath.com

WHY WE REMEMBER

WHY WE REMEMBER

WHY WE REMEMBER

UNLOCKING MEMORY'S
POWER TO HOLD ON TO
WHAT MATTERS

. . .

Charan Ranganath, PhD

VINTAGE BOOKS
A DIVISION OF PENGUIN RANDOM HOUSE LLC
NEW YORK

FIRST VINTAGE BOOKS EDITION 2025

Copyright © 2024 by Charan Ranganath

Penguin Random House values and supports copyright. Copyright fuels creativity, encourages diverse voices, promotes free speech, and creates a vibrant culture. Thank you for buying an authorized edition of this book and for complying with copyright laws by not reproducing, scanning, or distributing any part of it in any form without permission. You are supporting writers and allowing Penguin Random House to continue to publish books for every reader. Please note that no part of this book may be used or reproduced in any manner for the purpose of training artificial intelligence technologies or systems.

Published in the United States by Vintage Books, a division of Penguin Random House LLC, New York. Originally published in hardcover in the United States by Doubleday, a division of Penguin Random House LLC, New York, in 2024.

Vintage and colophon are registered trademarks of Penguin Random House LLC.

Some names and identifying details have been changed
to protect the privacy of individuals.

The Library of Congress has cataloged the Doubleday edition as follows:
Names: Ranganath, Charan, author.
Title: Why we remember : unlocking memory's power to hold on to what matters / Charan Ranganath, PhD.
Description: First edition. | New York : Doubleday, 2024. |
Includes bibliographical references and index.
Identifiers: LCCN 2023031694
Subjects: LCSH: Memory. | Cognition.
Classification: LCC BF371 .R287 2024 | DDC 153.1/2
LC record available at https://lccn.loc.gov/2023031694

Vintage Books Trade Paperback ISBN: 978-0-593-46783-1
eBook ISBN: 978-0-385-54864-9

Author photograph © Michael Rock
Book design by Betty Lew

vintagebooks.com

Printed in the United States of America
1st Printing

To my family

CONTENTS

WHY WE REMEMBER

WHY WE REMEMBER

MEET YOUR REMEMBERING SELF

. . .

> My ability to remember song lyrics from the
> eighties far exceeds my ability to remember
> why I walked into the kitchen.
>
> —Anonymous internet meme

Take a moment to think about who you are right now.

Think about your closest relationships, your job, your geographic location, your current life circumstances. What would you consider your most indelible life experiences—the ones that have made you who you are? What are your most deeply held beliefs? What choices, large and small, good and bad, have led you to this place, to this moment in time?

These choices are routinely influenced, and sometimes completely determined, by memory. To paraphrase Nobel-winning psychologist Danny Kahneman, your "experiencing self" does the living, but your "remembering self" makes the choices. Sometimes these choices are small and mundane, such as what to eat for lunch or which brand of laundry detergent you grab from a crowded supermarket shelf. Other times, they are the driving force behind life-changing decisions, from what career to pursue and where to live to what causes you believe in, even how you raise your children and what sort of people you want around you. And further, memory shapes the way you feel about those choices. Kahneman and others have shown in many studies that the happiness and satisfaction you gain from the outcomes of your decisions *do not come from what you experienced, but rather from what you remember.*

In short, your remembering self is constantly—and profoundly—shaping your present and your future, by influencing just about every decision you make. That's not necessarily a bad thing, but it does mean we need to understand the remembering self, and the mechanisms of its far-reaching influence.

Yet the pervasive involvement of memory in our thoughts, actions, emotions, and decisions often passes unnoticed, except for those moments when it fails us. I know this because, whenever I meet someone new and they discover I study memory for a living, the most common question I get is, "Why am I so forgetful?" I often ask myself the same question. Daily, I forget names, faces, conversations, even what I'm supposed to be doing at any given moment. We all wring our hands over those moments that we can't remember, and as we get older, forgetting can be downright scary.

Severe memory loss is undoubtedly debilitating, but our most typical complaints and worries around everyday forgetting are largely driven by deeply rooted misconceptions. Contrary to popular belief, the most important message to come from the science of memory is not that you can or even should remember *more*. The problem isn't your memory, it's that we have the wrong expectations for what memory is for in the first place.

We are not supposed to remember everything from our past. The mechanisms of memory were not cobbled together to help us remember the name of that guy we met at that thing. To quote British psychologist Sir Frederic Bartlett, one of the most important figures in the history of memory research, "literal recall is extraordinarily unimportant."

So instead of asking "Why do we forget?" we should really be asking, *"Why do we remember?"*

My first step toward answering this question began on a windy fall afternoon in 1993. I was a twenty-two-year-old graduate student working on my PhD in clinical psychology at Northwestern University, and I had just designed my first research study on memory—although it wasn't supposed to be about memory. I was doing research on clinical depression, and we designed the study to test a theory of how being in a sad mood affects attention. I walked into Cresap Laboratory with a song by Hüsker Dü (whose band name is Norwegian for "Do you remember?") blasting in my headphones to psych myself up as I prepared to collect electroencephalography (aka EEG) from my first subject. I found myself struggling to attach electrodes to the scalp of a college student with a head of thick curls. After thirty minutes of staring at the computer monitor, mesmerized by the waves of electrical activity emanating from her brain, it was time to remove the electrodes and clean up. Despite my best efforts, when she left the lab, her hair was spackled with a crust of thick conductive paste.

The idea was to make otherwise emotionally healthy subjects feel sad and then observe whether being in a sad mood led their attention to be captured by negative words (such as *trauma* or *misery*) more than neutral words (such as *banana* or *door*). To get our volunteers into a sad mood, we had them listen to a selection of slowed-down classical music, including "Russia Under the Mongolian Yoke," by Sergei Prokofiev, from the film *Alexander Nevsky*—a song so effective at inducing sadness it's been used in a number of studies on clinical depression. While the music played in the background, we asked our volunteers to reflect on a past event or time in their lives when they were sad. We expected that the music would make it easier to remember a sad event, and that remembering a sad event would make people feel sad. We were correct. Worked every time.

The rest of the experiment was a bust, but what stuck with me was that we were able to use people's memories of the past to change how they felt and looked at things in the present. It wasn't just that thinking about a painful event in their past made them sad; it seemed that *being* sad made it easier for them to remember other sad events. From that moment on, I became fascinated with how the structures of the brain that give rise to what we think of as "remembering" can profoundly affect how we think and feel in the present moment, and thus how we move into the future.

Memories can be triggered in a lab by a mournful piece of classical music, but in the real world they often sneak up on us at the most unexpected times and from the most unlikely sources—a word, a face, a certain smell or taste. For me, it only takes two chords from "Born in the U.S.A." to bring back a flood of memories about the people in junior high who regularly subjected me to all sorts of racist epithets.

The sounds, smells, and sights we experience in the present can also transport us back to joyful times. A song by the indie rock band fIREHOSE always takes me back to my first date with my future wife, Nicole; the smell of jackfruit reminds me of a walk on the beach with my grandfather in Madras, India; and the sight of the brightly colored mural outside a small Berkeley pub called the Starry Plough will send me back to my college days, when I played a memorable show with my college rock band, Plug-In Drug. (Yes, I regret that band name.)

Each of these remembered experiences and the feelings they elicit speak to one of the core principles that has underpinned much of my work, both as a clinical psychologist and a neuroscientist: Memory is much, much more than an archive of the past; it is the prism through which we see ourselves, others, and the world. It's the connective tissue underlying what we say, think, and do. My own career choices were, no doubt, influenced by my experiences as a first-generation immigrant that left me with a lasting sense of "otherness." So much so, I sometimes feel like an alien, probing human brains to try to figure out how and why people behave the way they do.

To fully appreciate the weird and wonderful ways in which the human brain captures the past, we need to ask the deeper questions of *why* and *how* memory shapes our lives. The various mechanisms that contribute to memory have evolved to meet the challenges of survival. Our ancestors needed to prioritize information that could help them prepare for the future. They had to remember which berries were poisonous, which people were most likely to help or betray them, which place had a soft evening breeze or fresh drinking water, and which river was infested with crocodiles. These memories helped them stay alive for their next meal.

Viewed through this lens, it is apparent that what we often see as the flaws of memory are also its features. We forget because we need to prioritize what is important so we can rapidly deploy that information when we need it. Our memories are malleable and sometimes inaccurate because our brains were designed to navigate a world that is constantly changing: A place that was once a prime foraging site might now be a barren wasteland. A person we once trusted might turn out to pose a threat. Human memory needed to be flexible and to adapt to context more than it needed to be static and photographically accurate.

This is therefore not a book about "how to remember everything." Rather, in the chapters to come, I will take you into the depths of your memory processes so you can understand how your remembering self can influence your relationships, choices, and identity, as well as the social world you inhabit. When you recognize the vast reach of the remembering self, you can focus on remembering the things you want to hold on to and use your past to navigate the future.

In part 1 of this book, I will introduce you to the fundamental mechanisms of memory, the principles behind why we forget, and how to remember the things that matter. But that is only the beginning of the journey. In part 2, we will dive progressively deeper into the hidden forces of memory that determine how we interpret the past and shape our perceptions of the present. Lastly, in part 3, we will explore how the malleable nature of memory allows us to adapt to a changing world and consider the larger implications of how our own memories are intertwined with those of others.

Along the way, you will get to know people whose lives have been dramatically affected by the idiosyncrasies of memory: some who remember too much and others who can't form new memories; people who are tormented by their memories of the past and those who have suffered greatly because of the memory errors of others. Their stories, and the more ordinary stories of people such as me, illustrate the (sometimes) invisible hand of memory that guides our lives.

Memory is more than just who we were, it's who we are and what we have the potential to become, as individuals and as a society. The story of why we remember is the story of humanity. And that story begins with the neural connections that seamlessly link our past to the present and our present to the future.

THE FUNDAMENTALS OF MEMORY

. . .

WHERE IS MY MIND?

WHY WE REMEMBER SOME THINGS
AND FORGET OTHERS.

. . .

> Maybe the reason my memory is so bad is that I
> always do at least two things at once. It's easier to
> forget something you only half did or quarter did.
>
> —Andy Warhol

Over your lifetime, you will be exposed to far more information than any organism could possibly store. According to one estimate, the average American is exposed to thirty-four gigabytes (or 11.8 hours' worth) of information a day. With a near-constant stream of images, words, and sounds coming at us through our smartphones, the internet, books, radio, television, email, and social media, not to mention the countless experiences we have as we move through the physical world, it's not surprising we don't remember everything. On the contrary, it's amazing that we remember anything. To forget is to be human. Yet, forgetting is one of the most puzzling and frustrating aspects of the human experience.

So it's natural to ask, "Why do we remember some events and forget others?"

Not long ago, Nicole and I celebrated the thirtieth anniversary of the year we met. To mark the occasion, we pulled out old family videos that had been gathering dust over the years and had them digitized. I was particularly fascinated by the footage of our daughter Mira's birthday parties. As we watched the videos of Mira growing up, I expected them to trigger a flood of memories. Instead, I discovered that almost all of them seemed new to me. I was the one shooting the videos, yet I did not have the experience of recollecting these parties as individual events—except for one.

For most of her early childhood, we organized Mira's birthday parties at such places as the Sacramento Zoo, the local science museum, a gymnastics studio, or an indoor rock-climbing gym. These venues ensure that the kids can be entertained and contained, with a steady stream of food, sugary drinks, and activities provided during the two-hour reserved window. At these birthday parties, I would participate in the festivities, but for the most part I focused on documenting these precious moments so that Nicole and I could revisit them later.

The year Mira turned eight, I decided to try something different. When I was a kid, my brother, Ravi, and I celebrated our birthdays at home. We had a lot of fun, and our parents didn't need to spend a lot of money. So, that year I followed my do-it-yourself punk-rock ethos and organized Mira's party at our house. Anyone who has ever hosted a children's birthday party knows the number one goal is to keep the kids busy. Mira was always into art, so I found a shop in a nearby town that provided premade cat-shaped ceramics that the kids could paint with glaze and later have fired to take home. Between the craft activity and the SpongeBob SquarePants piñata I had hung up in the backyard, I figured I had it covered.

I couldn't have been more wrong. Roughly fifteen minutes into the activity, all the cats were painted. With hours left to fill before cake time, the children were getting restless, and I was beginning to panic. I herded the kids out to the backyard, where they lined up to take turns whacking a piñata that refused to burst. Eventually, I took matters into my own hands, got out a golf club from the garage, and bashed a hole in it. Candy went flying everywhere and the kids descended on that papier-mâché SpongeBob like a scene from *The Walking Dead*. I saw one kid launch herself like an Olympic gymnast across the yard to get to a Snickers Mini she'd spotted in the grass.

It was still too early to bring out the cake, so I came up with the bright idea of having them play tug-of-war with an old rope I found in the garage. It had rained the day before, and the kids kept slipping and sliding around on the muddy grass. I remember looking around the backyard—some of the kids were chasing one another around in a sugar frenzy, one or two were complaining about rope burn, and a couple were taking turns beating the SpongeBob carcass to death with the golf club—and thinking how quickly an eight-year-old's birthday party can go from painting ceramics to *Lord of the Flies*. It was not my finest moment, but it *is* one I remember in excruciating detail.

Not all our experiences are of equal importance. Some are utterly unremarkable; others are moments we hope to treasure forever. Unfortunately, even priceless moments can sometimes slip through our fingers. At the time, I could have sworn I would vividly remember each of Mira's parties, so why is it that this one stands out and the other birthday videos seem like reruns from a distant TV show?

How can an experience that feels so memorable while we're living it ultimately be reduced to little more than a vague fragment of what transpired?

Although we tend to believe that we can and should remember anything we want, the reality is we are designed to forget, which is one of the most important lessons to be taken from the science of memory. As we will explore in this chapter, as long as we are mindful of how we remember and why we forget, we can make sure to create memories for our most important moments that will stick around.

MAKING THE RIGHT CONNECTIONS

The scientific study of memory as we know it today was pioneered in the late nineteenth century by German psychologist Hermann Ebbinghaus. A cautious and methodical researcher, Ebbinghaus concluded that, to understand memory, we must first be able to quantify it objectively. Rather than asking people subjective questions about events such as their kid's birthday parties, Ebbinghaus developed a new approach to quantify learning and forgetting. And unlike modern psychologists, who have the luxury of enlisting college students to voluntarily participate in their studies, poor Ebbinghaus worked alone. Like a mad scientist in a Gothic horror novel, he subjected himself to mind-numbing experiments, in which he memorized thousands of meaningless three-letter words called trigrams, each consisting of a vowel sandwiched between two consonants. His idea was that he could measure memory by counting the number of trigrams—e.g., DAX, REN, VAB—he was able to successfully learn and retain.

We should pause a moment to appreciate the painstaking work that went into Ebbinghaus's studies. In his 1885 treatise, *On Memory: A Contribution to Experimental Psychology*, he writes that he could only memorize sixty-four trigrams in each forty-five-minute session because "toward the end of this time exhaustion, headache, and other symptoms were often felt." In the end, his Herculean efforts bore fruit, as his experiments revealed some of the most fundamental aspects about the way we learn and forget. One of his most important achievements was to construct a *forgetting curve*, allowing him to graph, for the first time, how quickly we forget information. Ebbinghaus discovered that only twenty minutes after memorizing a list of trigrams, he had forgotten nearly half of them. One day later, he had forgotten about two-thirds of what he originally learned. Although there are some caveats to Ebbinghaus's findings, his bottom line holds: Much of what you are experiencing right now will be lost in less than a day. Why?

To answer this question, let's begin by breaking down how a memory is formed in the first place. Every area of the human *neocortex*, the densely folded mass of gray tissue on the outside of the brain, consists of massive populations of *neurons*—86 billion, according to one estimate. To put this number into perspective, that's more than ten times the human population of the earth. Neurons are the most basic working unit of the brain. These specialized cells are responsible for carrying messages to different areas of the brain about the sensory information we take in from the world. Everything we feel, see, hear, touch, and taste, every breath we take, every move we make (sorry, couldn't help myself), happens because of communication between neurons. If you feel yourself falling in love, if you're angry, or if you're slightly hungry, that's the outcome of neurons talking to one another. Neurons can also work in the background to handle important functions we're not even aware of, such as keeping our hearts pumping. They even work while we're sleeping, filling our heads with crazy dreams.

Neuroscientists are still working out exactly how all these neurons work together, but the knowledge we have so far is enough to build computer models that capture some of the basic principles that govern brain function. In essence, neurons function like a democracy. Just as one person has only one vote to influence the outcome of an election, a single neuron plays only a small role in any kind of neural computation. In a democracy, we form political alliances to advance our individual agendas, and neurons form similar alliances to get things done in the brain. The Canadian neuroscientist Donald Hebb, whose work was influential to our understanding of how neurons contribute to learning, called these alliances *cell assemblies*.

In neuroscience, as in politics, it's all about having the right connections.

To get a better sense of how this works, let's consider what happens as a newborn baby is exposed to human speech. Before a language is learned, babies can hear differences between sounds, but they don't know how to parse those sounds in a linguistically meaningful way. Fortunately, from the moment we are born, our brains get to work making sense of what we are hearing, trying to break up a continuous stream of sound waves into discrete syllables. What the baby ultimately perceives will depend on an election taking place in areas of the brain that process speech sounds. Perhaps the baby hears a sound, but there is some noise in the room, so it's unclear whether that sound was *bath* or *path*. Somewhere in the brain's speech centers a large coalition of neurons casts votes for the sound *bath*, a smaller coalition votes for *path*, and an even smaller minority votes for other candidates. Within less than half a second, the vote is tallied, and ultimately the baby picks up that it is time for a *bath*.

Here's where learning kicks in: In the aftermath of the election, the winning coalition works to strengthen its base. Neurons that only weakly supported the winning sound might need to be brought into the fold, and the ones that didn't need to be purged. The connections between the neurons that supported *bath* are strengthened, and connections with neurons that voted for the wrong sound are weakened. But at other times, the baby might hear someone loudly say the word *path*. The connections between the neurons that supported *path* will be strengthened and become less connected from the neurons that voted for the wrong word. Through these postelection shake-ups, the parties become more polarized; neurons will become even more strongly affiliated with the assemblies they already supported and pull further away from the ones they were lukewarm about. That leads the elections to become more and more efficient, so that the outcome of an election becomes apparent early in the voting.

Children's brains, in particular, are constantly in flux, reorganizing to optimize their perception of the environment. During their first few years, babies make dramatic progress at learning how to differentiate syllables, so that a continuous stream of sound can become sensible speech through the constant reorganization of connections between neurons. But as those neurons settle into coalitions that differentiate between the sounds the baby is hearing, they are becoming less sensitive to sound differences that don't exist in that language. It's as if the neurons are choosing between a small number of candidates based on a few key issues.

The baby's ability to change connections in the neocortex in response to new experiences is called *neural plasticity*. The reduction in neural plasticity as we transition to adulthood is well-known, though the science has been a bit distorted by news articles and TV shows conveying a bleak message that our capability for plasticity slips away as we get older. This message has been exploited by companies selling products that purport to stave off the inevitable decline. It's true that, past the age of twelve, the neural alliances formed around familiar sounds become more entrenched and it becomes harder to learn new kinds of syllables as quickly. This is why it can be more difficult to start learning Mandarin or Hindi in your forties than if you had been exposed to those languages as a child. Fortunately, adult brains still have plenty of plasticity without the need for any pills, powders, or supplements. The connections in your brain are constantly being reshaped with the goal of improving your perception, movement, and thinking as you gain more and more experiences. Moreover, as you go past simple perception (what we see, hear, touch, taste, and smell) and move into higher-order functions (e.g., judgment, evaluation, and problem-solving), the brain is remarkably plastic, and the neural elections are highly contested.

So, suppose you've spent a week in Delhi learning Hindi, and you'd like to ask for water at a restaurant. You memorized that word only an hour ago, but now you can't find it. Unfortunately, until you become more proficient, many Hindi words might sound alike to you. The cell assembly for the word you're looking for (*paani*) is not yet strongly connected, and a lot of neurons have divided loyalties, torn between competing possibilities. This is the same challenge we face when trying to remember more complex experiences such as my daughter's well-organized birthday party at the Sacramento Zoo. To get to what we want to remember, we must find our way to the right coalitions of neurons, but in many cases, there is an intense competition between the coalition that has the memory you're looking for and coalitions representing other memories you don't need at that moment. Sometimes, the competition isn't so bad, but if you have a lot of coalitions representing similar memories, the battles can be intense, and there might not be any clear winner. In memory research, this competition between different memories is called *interference*, and interference is the culprit behind a lot of our everyday forgetting. The key to escaping interference is to form memories that can fight off the competition, and fortunately, we have the capability to make that happen.

ATTENTION AND INTENTION

Let's imagine an everyday scenario. You come home from work, checking email on your phone as you put your key in the lock and open the front door. As you step inside, your exuberant, poorly trained, recently adopted rescue dog jumps all over you, leaving you wet with dog slobber. You hear loud music pumping from your daughter's room and a horribly catchy eighties synthesizer-heavy pop nugget burrows into your brain. You wearily walk into the kitchen, where a rancid odor lets you know you forgot to take out the trash the night before. Then, a twinge of pain reminds you that you need to ice that ankle you sprained a few weeks ago.

Now, without looking back, try to recall where you left those keys. If you remember leaving them in the lock, that's great, but if you have trouble remembering that, you're not alone. You were probably distracted by a lot of other stuff. When we face an onslaught of information, our memory for an event becomes cluttered. What's worse, when we try to remember where we last put our keys, we are sifting through memories of all the previous places we put our keys, and all the various circumstances in which we did it, whether it was last night, last week, or last year. That's a lot of interference. And that is why the things we lose track of so often—keys, phone, glasses, wallet, even our cars—are also the things we use frequently. Given all that competition, how do we ever manage to remember these things?

Think of memory like a desk cluttered with crumpled-up scraps of paper. If you'd scribbled your online banking password on one of those scraps of paper, it will take a good deal of effort and luck to find it. This is not unlike the challenge of remembering. If we have experiences that are, more or less, the same—like the meaningless trigrams Ebbinghaus struggled to memorize—it becomes exponentially harder to find the right memory when we need it. But if your password is written on a hot-pink Post-it note, it will stand out among all the other notes on your desk and you can find it pretty easily. Memory works the same way. The experiences that are the most distinctive are the easiest to remember because they stand out relative to everything else.

So how do we make memories that stand out in our cluttered minds? The answer: *attention* and *intention*. Attention is our brain's way of prioritizing what we are seeing, hearing, and thinking about. At any given moment, we could be paying attention to a multitude of things going on around us. All too often, our attention is grabbed by what is in our environment. In the imagined scenario I described earlier, you might briefly have focused on your keys before your attention was captured by what you were confronted with after opening the door. Even if you pay attention to the most important thing to remember (i.e., those keys you're going to need in an hour when you realize you're running late to pick up your partner at the airport), that doesn't necessarily help you make a distinctive memory that will overcome all the interference from everything else that captured your attention (the exuberant dog, the funky garbage smell in the kitchen, and the sound of Kajagoogoo emanating from your daughter's bedroom).

This is where intention comes in. To create a memory that you can locate later on, you need to use intention to guide your attention to lock on to something specific. The next time you put down an object you frequently lose track of, such as your keys, take a moment to focus on something that is unique to that specific time and place, such as the color of the countertop, or the stack of unopened mail next to the keys. With a little mindful intention, we can combat our brain's natural inclination to tune out the things we do routinely and build more distinctive memories that have a fighting chance against all the interfering clamor.

THE CENTRAL EXECUTIVE

For the most part, as we go about our daily lives, we do a pretty good job of focusing on what's relevant. For that, you can thank a part of the brain that sits just behind your forehead called the *prefrontal cortex*. The prefrontal cortex will come up many times in this book because it plays a starring role in many of our day-to-day memory successes and failures, and one of its many functions is to help us learn with intention.

The prefrontal cortex takes up about a third of the real estate in the human brain, yet it has been misunderstood for much of the history of neuroscience. In the 1960s, neurosurgeons routinely removed the prefrontal cortex as a treatment for schizophrenia, depression, epilepsy, and any forms of antisocial behavior. This brutal procedure, known as a frontal lobotomy, was often performed by administering local anesthesia and sticking an icepick-like surgical instrument behind a patient's eyeballs and basically wiggling it around to damage a large chunk of the prefrontal cortex. The whole procedure could be done in about ten minutes. After a successful lobotomy—and many were unsuccessful, causing serious complications and sometimes death—patients would walk and talk normally and did not seem to have amnesia, but they were calmer and more compliant, as if they had been "cured." In fact, rather than treating any underlying mental illness, the frontal lobotomy left patients in a zombielike state, apathetic, docile, and devoid of motivation.

Around the same time, a small but dedicated group of neuroscientists who were studying the prefrontal cortex (which is part of a larger region called the frontal lobes) began to appreciate the importance of this area of the brain. They could see that damage to the prefrontal cortex caused deficits in thinking and learning, but its function seemed mysterious. Papers from the 1960s through the 1980s stressed the enigmatic nature of this region with titles such as "The Riddle of Frontal Lobe Function in Man," "The Problem of the Frontal Lobe," and "The Frontal Lobes: Uncharted Provinces of the Brain."

The prefrontal cortex doesn't quite get the credit it deserves when it comes to human memory. If you've read any books or popular press on memory, you've probably been introduced to the *hippocampus*. This seahorse-shaped area tucked away in the middle of your brain is regarded as *the* key area that determines whether you will remember or forget something. It is true this area of the brain plays an essential role in memory, one I'll describe in the next chapter. But even though the hippocampus is the belle of the ball for most neuroscientists, the prefrontal cortex has a special place in my heart. It was my entry point into memory research, and it plays a key role in determining what is retained and what gets lost.

The textbooks used to tell us that the prefrontal cortex and the hippocampus are two different kinds of memory systems in the brain. The prefrontal cortex was seen as a "working memory" system, temporarily keeping information online, like the RAM on our computers, whereas the hippocampus was believed to be a "long-term memory" system enabling us to store memories more or less permanently, like a hard drive. The working memory system was envisioned by some neuroscientists as a kind of sorting station that houses the information we take in until it's either dumped or dispatched to the hippocampus to be packaged into a long-term memory. As we would soon learn, this view was overly simplistic and failed to capture the broad reach of the prefrontal cortex in all aspects of cognition.

By the midnineties, researchers began to use brain imaging techniques to uncover how brain areas such as the prefrontal cortex contributed to working memory. One imaging technique, positron-emission tomography, or PET, identifies areas of high blood flow in the brain by injecting people with water containing a radioactive tracer while they lie in a scanner equipped with sensors that detect the radioactive emissions. Early research showed that blood flow in the brain was increased around areas that were working hard and needed a lot of glucose to keep them going. Scientists could use this information to map the brain by scanning people while they did tasks emphasizing different abilities, such as perception, language, and memory.

Because it's expensive and generally better to avoid injecting people with radioactive tracers, PET soon became supplanted by a technique called functional magnetic resonance imaging, or fMRI, in which researchers could measure magnetic field changes produced by blood flow (thanks to hemoglobin, the iron-containing molecule that becomes sensitive to magnetic fields when it is not carrying oxygen).

In a typical fMRI study, a subject will lie in the bore of the magnet, on a flat bed in a tube with a magnetic field strength of 1.5 or 3 tesla (which is thirty thousand to sixty thousand times the strength of the earth's magnetic field), the person's head in a helmetlike coil that is used to scan the brain. The head coil has a mirror, angled so participants can look up and see a video screen, and they are given a box with buttons to push to respond during the experiment. Participants wear earplugs, because when the fMRI data is being collected, the scanner makes a loud, constant beeping sound. I know this sounds terribly unpleasant, but for me it's not; if anything, I find it easy to fall asleep in there.

To study working memory with fMRI, researchers might show a stream of numbers to the volunteer in the scanner, who has to keep the last number on the screen in mind. Each time a new number is shown, the volunteer has to decide if this number matches the last one shown. The test requires working memory because the volunteer has to keep in mind only the last number shown, then dump it in favor of the new number in anticipation of the next number. In variations of the test, experimenters had volunteers keep in mind the last two numbers, and so on. The more numbers people had to keep in mind, the more activity was apparent in the prefrontal cortex. This seemed like good evidence that the prefrontal cortex played a role in temporarily holding information.

When I was in graduate school at Northwestern University, I was fascinated by this research, but it didn't seem to connect with what I was seeing in the clinic at Evanston Hospital, where I trained in neuropsychology. Many of the patients coming in had been referred by their doctors, who suspected they had some form of brain damage. My job was to administer a series of cognitive tests to inform diagnosis and treatment. Some patients had problems with language (aphasia) or intentional movement (apraxia) or recognizing objects or faces (agnosia). Others had memory problems (amnesia), such as what happens in the early stages of Alzheimer's disease, epilepsy, or conditions that cause a brief loss of oxygen to the brain. These syndromes were easy to spot. But then there were the people with damage to the prefrontal cortex.

Sometimes there was an obvious trauma, such as the prosecutor who had suffered a stroke, the construction worker who got hit in the head with a girder, or the bus driver who had surgery to remove a brain tumor. Or a patient could have something like multiple sclerosis, in which the immune system runs amok, attacking the integrity of neural connections in the prefrontal cortex, as well as elsewhere in the brain. The common complaint from every one of these patients was that they were having problems with memory. But when I tested them, they didn't seem to have a memory deficit. Something else was going on. They could effortlessly hold a string of numbers in mind and recall them back to me, and they were fine on a task that emulated the electronic game Simon, in which they had to watch me tap a series of blocks and then tap the same blocks in the same order. In other words, these patients could keep information in working memory. They struggled, however, with tests that required them to focus in the face of distraction. For instance, we might ask them to keep in mind some numbers in the center of a screen while ignoring numbers that were flashed on the left or right side. In these instances, the patients tended to get distracted by the numbers on the sides and lost track of the ones in the center.

The patients with frontal dysfunction also performed inconsistently on tests of long-term memory, in which we asked them to memorize a long list of words such as *cinnamon* and *nutmeg*. If we simply asked them to *recall* the words without giving them any additional cues, the patients could only remember a few words. But if we asked them whether a specific word was on the list, they could easily *recognize* whether it had been on the study list. These patients formed memories for those words, but could not find those memories without highly specific cues. One reason they had such difficulty retrieving these memories is that they did not use any kind of memorization strategy and instead focused on whatever grabbed their attention at the moment. In contrast, healthy individuals typically used strategies that would help them perform well on both the recall and recognition tests (e.g., focusing on many of the words being names of spices).

After testing many patients, I found that people without a functioning prefrontal cortex could do fine when they were given clear instructions and no distractions, but they struggled when they had to spontaneously use memory strategies or follow through on a task when irrelevant things competed for their attention. These observations convinced me that, even though the prefrontal cortex doesn't "do" memory, damage to the prefrontal cortex affects people's memory in the real world.

After completing my clinical training in 1999, I moved into full-time research with Dr. Mark D'Esposito at the University of Pennsylvania School of Medicine. Mark was pushing the envelope to develop newer and better fMRI techniques to study working memory. But unlike most other cognitive neuroscientists, he split his time between the lab and the clinic, where he was an attending behavioral neurologist. Mark was keenly aware of the disconnect between the way many neuroscientists talked about the prefrontal cortex and the problems he was seeing in patients with damage to it. One of his patients, a truck driver named Jim, was unable to work or even live independently after a stroke left him with extensive frontal damage. Jim's wife explained he was having memory problems. He would forget entire scenes after watching a movie and would end up watching it again two or three times back-to-back. Or he would forget to brush his teeth or shave, whereas before the injury he was more fastidious. Beneath the memory issues, though, something else seemed to be going on. It wasn't that he had forgotten *how* to perform these activities—he was fully capable of brushing his teeth—but left on his own, he just didn't take the initiative to do it, or he would get distracted and move on to something else. Jim was not unlike the people I had tested at Evanston Hospital who did not use any kind of strategy to memorize words.

A number of us who were working in Mark's lab had been running fMRI studies of working memory, and our results consistently supported the idea that areas in the *back* of the brain had cell assemblies that seemed to store memories for specific kinds of information. One area might activate when someone was asked to keep a mental picture of a person's face in mind, and another area might activate when someone was asked to keep a mental picture of a house in mind. Activity in the prefrontal cortex was not particularly sensitive to what someone was keeping in mind, or even whether the person had to do a working memory task at all. But the prefrontal cortex was intensely activated when a person had to use intention to stay on task, focus on distinctive information, resist distractions, or initiate some kind of mnemonic strategy.

The studies we were doing on the prefrontal cortex bridged the gap between what was discussed in scientific papers and what we saw in the clinic. The textbook view—that the brain is composed of specialized memory systems, each matching up with a different kind of task—was missing the bigger picture. The prefrontal cortex is not uniquely specialized for any specific kind of memory. Instead, fMRI studies and observations of patients supported a different theory, in which the prefrontal cortex is the "central executive" of the brain.

The best way to understand this theory is to think of the brain as a large company. In a major corporation, you have a bunch of specialized divisions: engineering, accounting, marketing, sales, and so forth. The job of the CEO—the chief executive officer—is not to be a specialist but to lead the company by coordinating activities across all these divisions so that everyone is working toward a common goal. Likewise, several regions all over the human brain have relatively specialized functions, and the job of the prefrontal cortex is to serve as a central executive, coordinating activity across these networks in the service of a mutual aim.

After a frontal lobotomy or frontal damage from a stroke, specialized brain networks remain, but they are no longer working together in the service of an internal goal. Individuals with damage to the prefrontal cortex can appear perfectly normal if they are asked to do a specific memory task with clear instructions in a distraction-free environment. But without a prefrontal cortex, they are unable to use their intentions to learn on their own, nor are they able to effectively use what they remember to get things done in the real world. They might go to the supermarket to buy milk and get distracted by an elaborate display of potato chips. Or they might know about an upcoming doctor's appointment but fail to use any strategies (such as setting a reminder on their phone) to make sure that they don't forget it.

THE CARE AND FEEDING OF YOUR PREFRONTAL CORTEX

I am fascinated with the prefrontal cortex partly because the memory struggles of patients with frontal damage are directly relevant to the kinds of memory issues that challenge many of us in daily life. Even in the absence of physical damage, many factors can affect the functioning of the prefrontal cortex, which can cause significant memory problems. For instance, many of the patients I tested at the neuropsychology clinic in Evanston were referred to us for evaluation for possible Alzheimer's disease, but on further examination, they turned out to be clinically depressed. In older adults, depression can look a lot like the early stage of Alzheimer's, as in the case of a recently retired schoolteacher I once tested. He had always prided himself on being sharp but was now having a hard time focusing and kept forgetting things. MRIs revealed no obvious brain damage, but his cognition was not much better than someone with damage to the prefrontal cortex. It didn't occur to him or to his physician that these cognitive problems might be related to his having just gone through a divorce and his living alone for the first time in decades.

The prefrontal cortex is one of the last areas of the brain to mature, continually fine-tuning its connections with the rest of the brain throughout adolescence. So even though children can learn fast, they're not so great at focusing on what's relevant because they're easily distracted. This is even more of an issue for children with ADHD (attention deficit hyperactivity disorder), who struggle in school not due to a lack of comprehension but because they have difficulty paying attention in class, developing effective study habits, and using strategies to help them perform well on tests. Considerable evidence suggests that ADHD is associated with atypical activity in the prefrontal cortex.

The prefrontal cortex is also one of the first areas to decline as we transition into old age, and consequently we feel more forgetful. Fortunately, for most older adults the problem isn't our ability to form memories, it's that changes in our ability to focus our attention lead to changes in *how* we remember an event. For instance, you might have had the experience of failing to remember the name of someone you met at your cousin's wedding, even though you can remember all sorts of other random information from the encounter—that he had freckles or was wearing a bright yellow bow tie or couldn't stop talking about a recent trip to Nashville.

This tendency to recall the inane at the expense of the important increases as we age. Numerous studies have shown that older adults are worse than younger adults at remembering things when they are required to pay attention and ignore distractions, but they can be as good as, or occasionally even *better* than, younger adults at remembering the distracting information. As we get older, we can still learn, but we have more trouble focusing on the details we want to take in, and we often end up learning things that might be irrelevant.

Regardless of age, there is no shortage of factors that can make you feel as if your prefrontal cortex is fried. In the modern world, multitasking is probably the most common culprit. Our conversations, activities, and meetings are routinely interrupted by text messages and phone calls, and we often compound the problem by splitting our attention between multiple goals. Even neuroscientists aren't immune to multitasking—nowadays, at virtually every academic talk, you'll find scientists in the audience, myself included, with laptops out as they alternate between listening to the talk and responding to emails. Many people even pride themselves on their ability to multitask, but doing two things at once almost always has a cost. The prefrontal cortex helps us focus on what we need to do to achieve our goals, but that wonderful ability gets swamped if we rapidly shift back and forth between different goals. Indeed, cognitive neuroscientist Melina Uncapher of UC San Francisco and her colleagues have shown that "media multitasking"—toggling between different media streams such as text messages and email—impairs memory. Moreover, certain parts of the prefrontal cortex are thinned out, on average, in people who do heavy media multitasking. More research is necessary to understand whether frontal dysfunction is a cause or a consequence of multitasking, but either way, the message is the same. As my friend, occasional bandmate, and one of the world's leading experts on the prefrontal cortex, MIT professor Earl Miller, likes to say, "There is no such thing as multitasking; you just end up alternating between doing different tasks badly."

Several health conditions can also compromise prefrontal function. Hypertension and diabetes, for instance, can cause damage to the white matter in the brain—the fiber pathways that enable areas in the brain to communicate with one another. My colleagues and I have found that age-related white-matter damage seems to isolate the prefrontal cortex from the rest of the brain—imagine that the CEO is locked alone in a room with no phone or internet access. Infections might also have similar effects through inflammatory processes that manifest in the brain. For instance, people infected with COVID-19 early in the pandemic suffered a loss in executive functions such as attention and memory, along with changes in brain structure in some parts of the prefrontal cortex. Changes in prefrontal function may account for the "brain fog" (otherwise known as long COVID) reported by those who have been infected for an extended period, as well as by those with other infection-related disorders such as chronic fatigue syndrome.

The ways in which we neglect our mental and physical health can temporarily wipe out your prefrontal cortex. For instance, sleep deprivation can have devastating effects on the prefrontal cortex and memory. Alcohol also adversely affects the prefrontal cortex, and some research suggests that these effects can last for days after a drinking binge. As we'll explore in later chapters, stress can also zap prefrontal function. If you stay up all night drinking and doomscrolling internet news sites after a stressful week at work, don't be surprised if you spend the weekend battling brain fog.

Fortunately, we can do some things to improve the functioning of the prefrontal cortex, though they're not necessarily the things you might think. Your brain is a part of your body, so anything you can do to care for your body is also good for your brain, and hence your memory. Sleep, exercise, and a healthy diet—all things that are good for your physical and mental health—are good for your prefrontal cortex as well. Aerobic exercise, such as running, increases the release of brain chemicals that promote plasticity, improves the vasculature that delivers energy and oxygen to the brain, and reduces inflammation and susceptibility to cerebrovascular disease and diabetes. In addition, exercise improves sleep and reduces stress, thereby mitigating two of the most significant factors that might otherwise sap our prefrontal resources. All these factors collectively can make a difference in maintaining memory functions during aging. One particularly impressive study that tracked memory performance in over twenty-nine thousand participants found that people whose lifestyle incorporated some of the factors described above were better at preserving their memory capabilities across a ten-year period.

MINDFUL MEMORIES

The selective nature of memory means our lives—the people we meet, the things we do, and the places we go—will inevitably be reduced to memories that capture only a small fraction of those experiences. Rather than fighting the selectivity of memory in a futile attempt to remember more, we can embrace that we are designed to forget and use intention to guide our attention so we can remember what matters.

Most of us know what it's like to struggle to remember the name of a person we've just met. It's amazing we ever succeed at this because there's nothing inherently meaningful about the connection between a name and a face. Strategies such as simply repeating the name can help a bit, but this approach is often insufficient because it fails to emphasize this connection. To succeed, you need to use intention to focus on the *right* information, so that the next time you see that face you'll have a cue that becomes your lifeline to remembering that person's name. For instance, if we met at a party, and you know your Greek mythology, you might link my name with Charon, the ferryman of the underworld who transports the souls of the dead across the River Styx. If you can find some aspect of my appearance that reminds you of Greece, mythology, and/or dead people, you'll be set to pull up my name whenever you see my face again. The point of these strategies is to intentionally create meaningful connections that allow us to find our way back to the memories we want to hold on to.

This brings me back to my daughter's birthday videos. As video cameras became smaller and more portable, we used them to document Mira's milestones. Unfortunately, those moments behind the camera came with a cost. During most of my daughter's birthday parties, my focus was on filming. Consequently, I have blurrier memories of those important moments than if I had put the camera down and allowed my brain to do what it does so well.

The problem isn't necessarily with the technology, but rather that we are filtering our experiences through the lens of a camera. When we take a photo or video, we tend to focus on aspects of an experience that enhance our memory for visual details, at the expense of those that immerse us in the event, such as sounds, smells, thoughts, and feelings. Mindlessly documenting events can lead us to disengage from the cues we need to form the kinds of distinctive memories that help us rise above interference.

Fortunately, taking pictures or video doesn't always have an adverse effect on memory. The optimal approach is to balance the needs of the experiencing self and the remembering self. With some mindful intention, cameras can work to our advantage to help shape or even curate the memories we can revisit later on. When I travel, I don't like to spend time incessantly taking staged portraits or photographing landscapes and tourist attractions because these activities detract from my experience. Instead, I like to take candid shots of people laughing, surprised, or engrossed, or of unusual highlights, such as an unintentionally humorous sign or gaudy statue. By documenting a few select, distinctive "moments," I free my mind to directly experience the trip and pay attention to what's happening around me. Looking back on these distinctive photos brings me back to the parts of the trip that I want to revisit, and conversely, many of the less enjoyable aspects of the trip, such as the crowds, lines, and traffic jams, are left in the blur.

Life is short. The transient nature of memory can make life seem much shorter. We tend to think of memory as something that allows us to hold on to the past, when in fact the human brain was designed to be more than simply an archive of our experiences (we'll learn how much more in later chapters). Forgetting isn't a failure of memory; it's a consequence of processes that allow our brains to prioritize information that helps us navigate and make sense of the world. We can play an active role in managing forgetting by making mindful choices in the present in order to curate a rich set of memories to take with us into the future.

2

TRAVELERS OF TIME AND SPACE

HOW REMEMBERING TAKES US BACK
TO A PLACE AND TIME.

. . .

I see that time travel really does already exist. . . .
It exists within the power of our own minds.

—The Flaming Lips

The flip side to the frustration of forgetting is that we can occasionally be pleasantly surprised when a memory that seemed long gone suddenly pops into our head, transporting us back to a particular place and time. This is not a quirk of the brain. We often think about memory as a record of *what* happened, but the human brain has the remarkable capability to link up the "what" with *where, when,* and *how.* This explains why the experience of remembering is so often accompanied by an ephemeral sense of pastness that's almost impossible to put into words. It's also why, if we are in the right place at the right time, lost memories seem to find *us*—as has happened to me often throughout my life.

My parents brought me to the United States when I was less than a year old. I have lived nearly my whole life in Northern California, but almost all my relatives live in India. When I was growing up, we traveled back there about every four years to visit my grandparents, aunts, uncles, and cousins. In my childhood and adolescence, I had many distinctive experiences during my visits to India, but upon returning to California, memories of those events would always fade, as if they were separated from me by the thousands of miles that stood between my home and my grandparents' house. Although my first words were in Tamil, my parents' native language, I can't speak more than a few phrases anymore (much to the chagrin of my paternal grandmother). It sometimes seems as if all those memories are locked away in a hidden compartment, forever out of reach. But when I'm in India, those memories are right there.

After a disorienting seventeen-hour flight, I emerge from the customs zone at Chennai International Airport into another world. From the moment I step outside the air-conditioned terminal, I experience a sensory assault. The air is thick with humidity, and the sweltering summer heat feels like a sauna—the sweat from my every pore does nothing to cool me off. I absorb the Technicolor brightness of the city, from the vibrant hues of the women's saris in the bustling markets to the colorfully painted trucks on the roads. The constantly changing flood of odors can be nauseating (passing an open sewer) or intoxicating (the sweet scent of tropical flowers, the sea air at the beach, and the thick smoke from wood-burning fires used by vendors roasting peanuts). The next morning as the sun creeps over the horizon, I wake up jet-lagged to the din of tropical birds echoing across the neighborhood. When I am in Chennai, with this cacophony of sounds, colors, and smells all around me, I can grab on to memories from past visits that escape me back home.

That sense of being in a particular time and place is called *context*, and it is critical for our day-to-day memory experiences. A great deal of everyday forgetting happens not because our memories have disappeared but because we can't find our way back to them. In the right context, however, memories that have seemed long gone can suddenly resurface back to the forefront of our recall.

Why is it that, in the right context, I can access dormant memories—including words and phrases in an otherwise foreign language—that seem unreachable to me back home? The answer lies in the way our brains lay down memories for events.

MENTAL TIME TRAVEL

For a good chunk of the twentieth century, the study of memory was dominated by *behaviorism*, a school of thought centered on the premise that memory can be reduced to simple associations between "stimuli" (sounds, odors, or visual cues) and "responses" (the actions we make in response to those stimuli) observed by an experimenter. Most of the research on learning during the heyday of behaviorism was done with animals. Whether it was a rat making its way out of a maze, a pigeon learning to peck for a reward, or a human struggling to memorize a boring list of trigrams, the idea was the same: learning is a simple process of forming associations. Any attempt to get at how people understand and consciously recollect past events was seen as an unscientific and pointless exercise. For behaviorists, understanding memory meant discovering the right equations to quantify how quickly associations are learned and forgotten under various conditions. Reading research papers from this era is about as much fun as a trip to the dentist's office (no offense to my dentist, who is actually very good).

Against this bleak backdrop Endel Tulving, an Estonian-born professor of psychology at the University of Toronto, stepped into the picture. Tulving loved to speculate, not just about what happens in experiments but what goes on in people's heads. In 1972, he broke with behaviorist theory in a groundbreaking chapter in which he dispensed with thinking of memory as a repository of simple associations and instead proposed that humans have two very different kinds of memory. He coined the term *episodic memory* to describe the kind of remembering that allows us to call back, and even reexperience, events from the past. Tulving proposed that episodic memory can be differentiated from *semantic memory,* our ability to recall facts or knowledge about the world, regardless of when and where that information was learned. Tulving's key insight was that, to remember an event (episodic memory), we need to mentally return to a specific place and time; but to have knowledge (semantic memory), we need to be able to use what we previously learned across a range of contexts.

By suggesting that memory is more than a gray lump of stimulus-response associations, Tulving completely eschewed the alluring simplicity of behaviorism. He later went so far as to call episodic memory a form of "mental time travel," meaning that remembering puts us in a state of consciousness in which we feel as if we were transported to the past. As Tulving put it, a key characteristic of human consciousness is that we are "capable of mental time travel, roaming at will over what has happened as readily as over what might happen, independently of the physical laws that govern the universe." When I first read this description, I thought Tulving had gone off the deep end—it didn't sound very science-y to be talking about time travel and consciousness. But with only a little bit of introspection one realizes Tulving was onto something.

Suppose I ask you to tell me some things you know about Paris. You might start off by saying that Paris is a city in France, home of the Eiffel Tower, and famous for its art museums and fancy restaurants. You would probably be 100 percent confident of those facts, even if you can't recall where or when you first learned them. Now, suppose I ask whether you have ever *been to* Paris. If you have, in answering that question you might pull up information that draws you back into a specific experience: the aroma of roasting chestnuts wafting from a street vendor's stall on the walk from your hotel to the metro; standing in line to take the elevator to the top of the Eiffel Tower on a chilly fall evening just before sunset; the view looking down on the city as daylight fades and the tower lights sparkle to life. It's not a matter of strong or weak memories—you can confidently pull up facts about Paris (semantic memory) and vividly reexperience a trip to Paris (episodic memory), yet the two experiences are totally different.

Initially, Tulving's proposal was divisive in the world of psychology. But, over the next fifty years, scientists would amass a body of scientific evidence validating Tulving's speculation that we have the ability to reboot our mind to the state it was in during a past event. Far more than simple recall, episodic memory connects us to those transient moments from the past that make us who we are in the present.

HUMANS 1, ROBOTS 0

The distinction between episodic and semantic memory is key to what makes humans fast and intelligent learners. One source of evidence for this point, ironically, comes from studies about the kind of learning that is challenging for machines to accomplish. Many of the most sophisticated applications of artificial intelligence, or AI, ranging from smart assistants such as Alexa and Siri to self-driving cars, are based on "neural networks"—computer programs that mimic, in an abstract way, how learning happens in the brain. Each time a neural network is trained to learn a fact, connections between simulated neurons in the network are modified. As the neural network is trained to learn more and more facts, the simulated cell assemblies in the model are constantly rearranged, no longer voting for any particular fact that was learned, but instead representing an entire category of knowledge. So, for instance, you might teach it:

"An eagle is a bird. It has feathers, wings, and a beak, and it flies."

"A crow is a bird. It has feathers, wings, and a beak, and it flies."

"A hawk is a bird. It has feathers, wings, and a beak, and it flies."

Eventually, the computer model becomes good at learning about new birds because it leverages what it already knows. If the network learns that a seagull is a bird, cell assemblies in the model can fill in the blanks and figure out that seagull can fly. But what if you teach it something a little different?

"A penguin is a bird. It has feathers, wings, and a beak, and it *swims*."

Now, the machine will have problems—the penguin fits all the characteristics of the bird except for one. A penguin is the exception to the rule that all birds fly, so when the computer learns the exception, it forgets what it previously learned about all the characteristics of birds. This problem is called *catastrophic forgetting*, and for machine learning, it's as bad as it sounds. The solution is to make sure the machine learns excruciatingly slowly, so that it doesn't immediately let go of the rule to learn the exception. This means it takes tons of training for neural networks to accomplish a task, making it difficult for them to adapt quickly to the complexity of the real world. Even today, the most sophisticated products of artificial intelligence must be trained on a massive amount of data to do anything interesting.

Humans, like the neural network models I described, are great at extracting general knowledge from past experiences, so that we can make assumptions and inferences about future situations ("That looks like a bird, so I can expect it could fly away"). But, unlike machines, we don't glitch every time we encounter variations in learning because we also have episodic memory. Episodic memory is not designed to capture the common elements of all our experiences; it stores and indexes every event differently, so you don't get mixed up when you learn the exception to the rule.

Armed with episodic *and* semantic memory, we can quickly learn both the rule (most birds fly) and the exceptions (penguins are birds that swim). In the real world, this enables us to pick up the information we can usually rely upon, such as the optimal route on our daily commute to work, while being flexible enough to adapt to unusual circumstances, such as taking an alternative route when we remember that the road is temporarily closed due to construction.

By pulling together information from neuroanatomy, studies of brain activity from neuroscience, studies of the effects of brain damage in humans, and computational models, scientists have generally reached the conclusion that the brain solves the problem of catastrophic forgetting by having systems that learn in different ways. The neocortex, the huge gray mass of brain tissue I described in chapter 1, works like a traditional neural network, enabling us to pick up facts, whether it's knowledge about birds or the typical weather in June in Chennai. The hippocampus, that area tucked securely in the middle of the brain I also mentioned in the previous chapter, is responsible for the brain's amazing ability to rapidly create new memories for events, so we can quickly learn the peculiar experiences that do not fit with our prior knowledge, such as a temperate and dry summer day in Chennai.

MEMORY CODES

The hippocampus may be the most intensely studied brain area in all of neuroscience. To many neuroscientists it's synonymous with memory, in part because of a study by Brenda Milner, a pioneering neuropsychologist. In 1957, she published the paper that introduced the world to Patient H.M.—as he was known in scientific literature to protect his identity. We now know he was Henry Molaison, a young man who suffered from debilitating seizures that had plagued him for more than a decade, preventing him from holding down a job or living a normal life.

In his late twenties, H.M. agreed to undergo a radical experimental surgery to treat his seizures by removing about five centimeters of tissue from the left and right hippocampus, along with the surrounding neocortical tissue in the *temporal lobes*. The surgery, performed by brain surgeon William Scoville, reduced the severity of H.M.'s epilepsy, but also caused him to become densely amnestic. H.M.'s memory disorder was so severe that you could start a conversation with him, then leave the room for less than a minute, and when you returned, he would have no memory that the conversation had taken place. Milner's paper, which definitively linked the formation of new memories to the hippocampus, was a shot heard around the world, inspiring an entire generation of scientists to understand why and how this little area in the human brain enables us to bring the past back to life. The impact of her contribution to the science of memory was so profound that, a few years after Milner published her study of H.M., the legendary Russian neuropsychologist Alexander Luria sent her a note in which he wrote, "Memory was the sleeping beauty of the brain and now she is awake."

After Milner's landmark publication, the question in neuroscience was not *whether* the hippocampus is important for memory but *how*. Further studies revealed that H.M. and other people with dense amnesia (due to various causes, such as herpes encephalitis or Korsakoff's syndrome) seemed to have equally severe problems with remembering recent events and learning new facts. This led some to conclude that the hippocampus must be an all-purpose memory device, and that, as far as the hippocampus goes, Tulving's distinction between episodic and semantic memory was irrelevant.

That conclusion was premature. Brenda Milner's original publication made it clear that H.M. had damage to areas outside the hippocampus. When MRI scanning technology became available, it soon became apparent that was an understatement. Scoville had removed about a third of H.M.'s temporal lobes and along the way tore through a significant chunk of white matter that would normally enable many otherwise intact brain areas to communicate with one another. As a result, we couldn't know what kind of memory functions were supported specifically by the hippocampus, as opposed to the many other brain areas that were affected by the surgery. To answer this question, we would need to study memory in people with damage that was much more specifically localized to the hippocampus.

In 1997, Dr. Faraneh Vargha-Khadem, a neuropsychologist at University College London, did just that, and discovered that Endel Tulving was right to distinguish between semantic and episodic memory. Faraneh had been studying teenagers and young adults with *developmental amnesia*, a term she coined to describe people suffering from memory problems at a very young age. Tragically, this happens more commonly than you might think; causes can range from premature birth, diabetic hypoglycemia, near-drowning incidents, or lack of oxygen to the brain at birth when the umbilical cord gets wrapped around the infant's neck. In all these cases, the hippocampus is the first region of the brain that is affected. In her groundbreaking 1997 report, Faraneh described three individuals, all of whom had experienced damage specifically to the hippocampus during childhood. Based on studies of H.M., one might expect that these children should have remained developmentally stunted, unable to gain the knowledge needed to navigate the world. In fact, though they had significant amnesia for events, they were able to acquire new semantic knowledge in school, though they probably learned more slowly than someone with an intact hippocampus.

That same year, Faraneh invited a group of scientists, including Endel Tulving, to London to interact with one of the individuals described in her paper, a teenager named Jon who was diagnosed with developmental amnesia at the age of eleven. Despite his amnesia, Jon had an impressive command of historical knowledge, easily reciting such facts as "At the time of World War I, the British Empire occupied about a third of the planet's landmass." Later that day, the scientists took Jon out to lunch—but Endel Tulving stayed behind to construct a memory test that he sprang on Jon when he returned. Tulving's questions revealed that Jon had almost no recollection of what had transpired at lunch, the route they took to the restaurant, or what he had seen along the way. The discrepancy between Jon's semantic and episodic memory was so great, Tulving remarked, that "he does not resemble any other kind of patient who has ever been described."

Research on patients like Jon has shown unequivocally that episodic memory depends on the hippocampus. Since then, fMRI studies have filled in the picture, giving us a window into how the hippocampus works in the intact brain. Significant advances in this area came about with the introduction of a new fMRI technique that allowed us to peer into the brain while someone recalled a specific memory, like that trip to Paris. This technique allowed us to move beyond merely looking at how the brain lights up, and instead to read out signals from individual events, so we could understand what makes each of our memories unique.

Here's how it works: If you look at fMRI data in the hippocampus while someone is doing a memory experiment, at any given moment some pixels will appear a little darker and some will appear brighter. The patterns change a little bit from moment to moment, so a pixel might become brighter or dimmer. We used to think of these moment-to-moment changes as "noise" produced by weird issues with MRI scanning, but it's now clear that some of this variation is meaningful. In 2009, I was having lunch with my friend Ken Norman, currently chair of the Department of Psychology at Princeton, who convinced me to look more closely at these patterns of brain activity in our memory studies. I began to wonder, What if every time we call up a memory for a particular event, a unique pattern of brain activity corresponds to that event? What if each pattern of light and dark pixels was like a QR code you might scan with your phone, with each unique configuration acting as a pointer to a particular memory? If so, we could use MRI scanning to read out "memory codes" that tell us how memories are sorted in different brain areas.

For instance, if I were to lie down in the MRI scanner and recall seeing my brother, Ravi, playing with his dog at a recent family picnic in the park, then recalled seeing him a few years ago, walking his dog on a grimy sidewalk in his San Francisco neighborhood, perhaps we would discover similar memory codes for each of those experiences. In our studies, this is exactly what we saw in areas of the neocortex that seemed to store the general facts about who (Ravi) and what (his dog, Ziggy) were in the event. In the hippocampus, however, the memory codes for those two events looked totally different. On the other hand, when we looked at the hippocampus while a person recalled two different pieces of information from the same event—such as seeing Ravi at the park versus seeing my wife, Nicole, at the same family picnic—the memory codes in the hippocampus were similar.

These findings helped to unlock the mystery of how the hippocampus helps us perform mental time travel. The cell assemblies that allow us to remember particular parts of an event—Ravi's face, the taste of the sandwiches at the picnic, the sound of his dog barking in the background—are in separate areas of the brain that normally do not talk to one another. The only thing these cell assemblies have in common is that they were active around the same time. The hippocampus, however, has connections to many of these areas, and its job is to store links to the different cell assemblies that come to life at a given moment. Later on, if I revisited that park, my hippocampus would help reactivate all those cell assemblies, enabling me to reexperience seeing Ravi. The hippocampus enables us to "index" memories for different events according to *when* and *where* they happened, not according to *what* happened.

The way the hippocampus forms memories has an interesting side benefit. Because the hippocampus organizes memories according to the context, recalling something from one event makes it easier to retrieve other events that happened around the same time or place, painting a fuller picture. Recalling the moment we cut a watermelon at the picnic would lead us to pull up the sequence of events that followed, such as playing Frisbee and volleyball a few minutes later. The hippocampus carries us back and forth in time, and we don't even need a wonky DeLorean to do it.

HERE AND NOW

What makes episodic memory such a powerful force is that it's not just for accessing the past. Part of our fundamental perception of reality is our ability in the present to orient ourselves in time and space, and we often have to recall the recent past to do so. Think of a time you woke up in the middle of the night in an unfamiliar bed and your first thought was "Where am I?" To help you answer this question, the hippocampus gets to work pulling up the right memory code; maybe you recall that, a few hours ago, you checked into a hotel, and with that information the moment of disorientation quickly passes. Pulling up a memory of the recent past helps to ground you in the here and now. According to one prominent theory, episodic memory emerged in evolution from the more basic ability to learn where we are in the world. That ability turns out to be crucial for survival, as I learned from a fortuitous collaboration with a young graduate student named Peter Cook.

I first met Peter at a memory conference. After several students presented research on how humans learn lists of words, Peter took the stage and played a series of short videos of his experiments on learning in California sea lions. His research captured my imagination—it had never occurred to me that one could even study memory in sea lions. Immediately after his talk, I introduced myself and talked Peter into inviting me and my family to visit his lab at the University of California, Santa Cruz. Mira, who was five, got to meet a sea lion up close, and she even helped out with data collection. Peter ran one of his memory tests while we were there, and Mira got to pull the levers that opened the doors and to press the buttons that played the sounds to cue the sea lions during each trial.

During our visit, I learned that Peter was studying the effects of domoic acid on the hippocampus. This marine biotoxin, which is produced during harmful algal blooms called red tides, works its way up the food chain as the algae are eaten by clams, mussels, and other shellfish, which are then eaten by crabs and fish, which are in turn consumed in large quantities by sea lions, who become exposed to high concentrations of domoic acid. Humans who ingest this toxin can get "amnesic shellfish poisoning," characterized by vomiting, nausea, confusion, and memory loss, and the same appeared to be true of sea lions exposed to domoic acid. Peter had the unique opportunity to put these sea lions in an MRI scanner, and he found that animals with domoic acid poisoning had significant damage to the hippocampus.

After this visit, Peter and I agreed to collaborate on what would become one of the most interesting brain-imaging projects of my career. I helped Peter develop new memory tests for the sea lions. In one of these tests, the sea lions had to remember the locations of fish that Peter had put into specific hiding places. Another test required them to keep track of what they had done recently in order to efficiently collect fish that were placed in different buckets. The sea lions with domoic acid poisoning performed terribly on these tests. We could predict how badly they would perform just by looking at the degree of damage to the hippocampus. Our research helped to explain why these poor animals were washing up onshore. Without a functioning hippocampus, they grew disoriented. Lost, and unable to recall their foraging sites, they became malnourished and ultimately stranded onshore.

When I saw Peter's results, it occurred to me that we often rely on episodic memory to orient ourselves in ways we don't even realize. Remember being in the hotel? Now, imagine waking up and having no context for what day it is or where you are, disoriented with nothing to anchor you in space and time. This is the tragic reality experienced by the millions of people suffering from Alzheimer's disease. The hippocampus is one of the first areas of the brain ravaged by Alzheimer's, and this is probably why patients in the early stages frequently get lost and lose track of the passage of time. A friend who is a caregiver for a parent with Alzheimer's shared with me how painful it was to see the look of fear on her mother's face when she became unmoored from her sense of when and where she was in the world. I imagine it must be terrifying, like treading water in the open ocean.

THE TIME MACHINE

Although the hippocampus enables us to mentally travel back to a place and time, I want to be clear that the brain has no direct way of perceiving our location or the exact time on a clock. It's not as if our memories have a time stamp or GPS coordinates telling us exactly when and where something happened. Rather, the hippocampus seems to track time by capturing changes in the world around us. We move around from place to place over the course of a day. Those places, ranging from small, enclosed rooms to the open sky of the outdoors, each have distinct sights, sounds, and smells, giving us a sense of "where" we are. Moreover, the environment around us is constantly changing. Day turns into night, satiety progresses into hunger, elation can transition to fatigue.

All these external factors, along with the motivations, thoughts, and feelings that characterize our internal world, come together to form the unique context that envelops our experience at any given time. When we access a particular episodic memory, we can pull up a bit of that past mental state, too, giving us the feeling of being back in that time and place. Changes in context over time, in turn, drive changes in our brain activity patterns that we experience as the passage of time. Two events that occurred close together in time—such as making coffee and eating breakfast—are going to have more contextual elements in common than events that occurred further apart in time, such as eating breakfast versus making dinner.

Context is such an integral component of episodic memories that it can have powerful effects on what we can remember. Being in a particular place, as when I am surrounded by the sights and sounds of my grandparents' homes in India, can bring back memories that otherwise elude us. Smells and tastes are another great cue. This was depicted quite effectively at the end of the Pixar film *Ratatouille* when a spoonful of the humble French stew transports a curmudgeonly food critic back to his childhood, when his mother had prepared a similar dish.

Music is yet another powerful cue for episodic memories. This is why a song you haven't heard since you were seventeen can transport you back to the high school dance where you had your first kiss. My UC Davis colleague Petr Janata has done studies cataloging the music people listened to during different time periods and finds that songs are exceptionally effective at eliciting mental time travel. Others have found that music can elicit recollections of past events in those with Alzheimer's disease. I saw this firsthand when my paternal grandfather, a filmmaker in South India, succumbed to dementia. Toward the end of his life, his memory had deteriorated and he sometimes had trouble recognizing me, but he could still sing the songs he had composed for his films, and these songs helped him pull up otherwise inaccessible memories from this period of his life.

Our emotions also contribute to context, which means that our feelings in the present affect what we can recall from the past. When we get angry, it's easy to pull up all those memories that give us more reasons to be annoyed, and it's harder to access the memories that don't. For example, you might have no trouble recalling positive memories about a romantic partner when things are going well, but it might be harder when you are arguing about whose turn it is to walk the dog or wash the dishes.

The central role of context in episodic memory sheds some light on why we forget and how we can overcome forgetting in the face of massive interference. As I mentioned in chapter 1, our most common (and frustrating) memory challenges come from repetitive experiences, such as trying to remember where you left your keys or whether you took your medicine in the morning. Consider the problem of finding your wallet. Did you leave it on the coffee table? On your desk at work? Or is it in your jacket pocket? At *some* point, your wallet has been in *all* these places, but that's not important—what you need to recall is the *last* place you put it. If the hippocampus only stored photographic memories of *what* happened, this task would be nearly impossible—you'd have a massive pile of "wallet" memories to sort through. Instead, the essential trick performed by the hippocampus is that it takes in information about the things in which we are interested—say, the wallet and the coffee table—and it ties it up with information about the context, all the other stuff that's going on in the background, such as the TV show that's on, the smell and taste of the coffee you sipped right after putting down the wallet, and the sensation that you feel hot and need to turn on the air conditioner. We experience zillions of repetitive events, but the context makes each unique. That means we can use context as a lifeline to find our way back to those things we seem to always lose.

When you're late for work and frantically searching for something, such as a lost wallet, especially if you're in a rush, you might start with a strategy based on semantic memory—by searching based on the knowledge of where you *usually* keep your wallet. But you can also tap into episodic memory to retrace your steps. Try to vividly remember where you were and what you were doing when you last recall having your wallet. If you can mentally time travel to the moment that you put your wallet down, the hippocampus can help you pull up other information from around the same time. The closer you can get to that context, the easier it will be to find the wallet.

Just as being in a particular place, situation, or state of mind makes it easier to access memories of other events that occurred in similar contexts, being in the wrong context can make it hard to find the right memory. Suppose you go to a party, and after a couple of glasses of wine, you find yourself deep in an animated conversation with a stranger. The next day, you run into her at the supermarket but can't quite place who she is or how you know each other. The problem is that the hippocampus didn't just store that person's face in your memory, it *connected* it to the context—the midcentury-modern furniture in the house, your slight buzz from that second glass of merlot, and the ambient noise of dance music and party guests talking. Without any of those context cues, it can be hard to bring yourself back to a conversation you struck up with someone while you were both waiting in line for the bathroom.

The further back in time you try to go, the harder it is for your brain to pull up a past context, and in some cases you won't be able to do it. Despite anecdotal claims to the contrary, scientific research has established that adults do not have reliable episodic memories from before the age of two. This phenomenon, known as infantile amnesia, is an enigma to scientists because very young children are fast learners and appear to be capable of forming episodic memories, but for some reason we can no longer access those experiences as we progress to adulthood. One possibility, based on research by my UC Davis colleague Simona Ghetti, is that the hippocampus is still developing during the first few years of life, so very young children lack the ability to link their experiences to specific spatial and temporal contexts. I also suspect that infantile amnesia happens because connections between neurons across the entire neocortex undergo massive reorganization during the first few years of development. It would be nearly impossible for an adult to travel back in time to infancy because our brains would have to undo years of wiring changes to get us back to the mental state we were in as babies.

WHAT WAS I LOOKING FOR AGAIN?

You've probably had the experience of walking into a room and having no memory of why you went there in the first place. This doesn't mean you have a memory problem—it's actually a normal consequence of what memory researchers call *event boundaries*. When you're in your home, you have a sense of where you are. If you step out the front door, that sense will dramatically change, even though you've only moved a short distance. We naturally update our sense of context when we experience a shift in our perception of the world around us, and those points mark the boundary between one event and another.

The context change that occurs with an event boundary has significant implications for episodic memory. Just as walls are physical boundaries dividing a house into separate rooms, event boundaries organize the timeline of our past experiences into manageable packets. People are better at remembering information that occurred at an event boundary than they are at remembering information from the middle of an event. Recent work from a number of labs suggests that this is because the hippocampus waits to store a memory for an event until right after an event boundary—that way, we only encode the memory once we have a full understanding of the event.

Given that our sense of context suddenly changes at event boundaries, it can sometimes be hard to recall things that occurred only moments earlier. At least once a week I find myself walking into the kitchen, scratching my head, and wondering, "What was I looking for again?" Inevitably, my frustration leads me to grab some junk food from the refrigerator, scarf it down, and walk back to my desk, only to realize as soon as I sit down that I went to the kitchen to retrieve my glasses. I've consumed many empty calories thanks to event boundaries.

Event boundaries happen all the time and don't necessarily require a change in location. Anything that alters your sense of the current context—a shift in the topic of conversation, a change in your immediate goals, or the onset of something surprising—can lead you to put up an event boundary. You've likely experienced this if you've ever been in the middle of telling a story and someone interrupted your train of thought, say, to point out your shoe was untied, and you forgot entirely what you were about to say. It can be frustrating—even alarming as we approach middle age and beyond—to have to ask, "What were we talking about?" But rest assured, it's a normal by-product of the way our brains use context to organize episodic memories.

Beyond causing these quirks in forgetting, event boundaries can also affect our sense of the passage of time. In 2020, millions of people all over the world endured months of lockdowns during the first wave of the coronavirus pandemic. The monotony of spending all day, every day in the same place, deprived of the usual activities that would normally provide structure to our daily lives, such as school schedules and work commutes, left many of us feeling as if we were no longer anchored in time and space. To get a sense of the time warp that people were experiencing, I polled the 120 students in my (online) Human Memory class about how they were experiencing the passage of time. After spending almost an entire semester stuck in the same room, staring at a computer screen, binge-watching television shows or attending online classes and lectures, an overwhelming majority (95 percent) said they felt that the days were going by slowly. Yet, without distinctive memories of what was happening during those days, most of them (80 percent) *also* felt that the weeks were passing by too quickly.

With few event boundaries to provide meaningful structure to their lives, my students—along with millions of other people all over the world—felt as if they were living in the twilight zone, floating aimlessly through time and space.

MAKING THE MOST OF MENTAL TIME TRAVEL

Nostalgia, that bittersweet mixture of joy and sadness that infuses so many of our most precious memories, is one of the most powerful ways episodic memory influences our everyday lives. On average, people find it easier to recall positive experiences than negative ones, and this positivity bias increases as we get older, which might explain older adults' penchant for nostalgia.

An abundance of research suggests reliving happy experiences

can improve our mood and self-confidence, and therefore our optimism about the future. That memory moment from *Ratatouille* I
referenced earlier in the chapter resonated so deeply with audiences
because we see ourselves in the crabby critic. This scene reminds us
how a simple context cue can transport us back to a happier time,
perhaps even changing our perspective and shifting how we see ourselves and our place in the world.

When we look back at the past, we tend to focus on a specific
period of our lives, between the ages of ten and thirty. The predomi-
nance of memories from these years is called the *reminiscence bump*,
and it isn't just apparent when we ask people to recall events from
their lives; it also shows up indirectly when people rattle off lists of
favorite movies, books, and music. Something about listening to a
song or watching a movie from those formative years can give us a
sense of meaning, connecting us to an idealized sense of who we are.

Although nostalgia can make us happy, it can also have the opposite effect, depending on the memories we choose to reflect upon
and the way in which we make sense of them. The term *nostalgia*
was coined by a Swiss physician in the late seventeenth century to
describe the particular kind of anxiety disorder he observed in mer-
cenary soldiers living far from home. For them, memories of good
times in a familiar place only highlighted their unhappiness in the
present. More recently, researchers found that, if people felt lonely
in their daily lives, engaging in nostalgia left them feeling even more
isolated and alone. In other words, the cost of nostalgia is that it can
leave us feeling disconnected from our lives in the present, giving us
a sense that things aren't as they were in the "good old days."

Rumination—repeatedly circling back to negative events and spin-
ning your wheels—is the evil twin of nostalgia, and a prime example
of how *not* to use episodic memory. People who have been identified
as having Highly Superior Autobiographical Memory because they
can pull up detailed recollections for seemingly trivial experiences
from the distant past tend to ruminate too much. As one such individual put it, "I do tend to dwell on things longer than the average
person, and when something painful does happen, like a breakup or
the loss of a family member, I don't forget those feelings."

To benefit from mental time travel, it's helpful to think about why the human brain evolved that capability in the first place: to learn from singular experiences. When we travel to past contexts, we can access experiences that reorient our view of the present. Recalling negative events can remind us of past lessons we have learned, so we can make better decisions in the present. Recalling positive events can help us to *be* better, by increasing altruism and compassion. In one study, people who vividly recalled an event in which they helped someone were more empathetic to the plight of others and expressed more willingness to help a person in need. By recalling past moments of compassion, wisdom, perseverance, or courage, we can use our connection with the past to broaden our sense of what we can do and who we can be.

3

REDUCE, REUSE, RECYCLE

HOW WE CAN REMEMBER MORE
BY MEMORIZING LESS.

. . . .

> We usually use the word *intuition*—sometimes also
> *judgment* or even *creativity*—to refer to this ability
> of experts to respond . . . almost instantaneously.
>
> —Herb Simon

As we've seen, we encounter far too much information in life to remember everything we experience or are exposed to. Fortunately, we don't have to. We can exploit what we know to organize our experiences, packaging countless bits of information into manageable chunks. With increasing experience, we can acquire expert skills that enable us to rapidly identify familiar patterns, helping us to remember the past, make sense of the present, and predict the future.

When we think of elite athletes, American Olympic gymnast Simone Biles, Jamaican sprinter Usain Bolt, or Argentine soccer player Lionel Messi come to mind. One who might seem out of place on that list is a former chemical engineer from Fayetteville, North Carolina. Yet, in the world of "memory sport," Scott Hagwood is a legend. A four-time winner of the USA Memory Championship, he is the first American memory athlete to be ranked by the International Association of Memory as a grand master.

Unlike elite professional athletes, who appear to be gifted with almost superhuman physical talents, Scott Hagwood was not born with any exceptional abilities. By his own account, he was an average student with mediocre grades and paralyzing test anxiety. Then, at the age of thirty-six, he was diagnosed with thyroid cancer and told that the radiation therapy necessary to save his life would also ravage his memory. Fearing that if his memory began to deteriorate, a fundamental part of his sense of self might be lost along with it, Scott turned to the science of memory to combat the side effects of his treatment.

After stumbling on a book by British memory trainer Tony Buzan, Scott began practicing a memory exercise using a pack of playing cards. He was stunned by the results and soon after won a bet with his brother that he could memorize a freshly shuffled deck in ten minutes. A year later, with his cancer in remission, he decided to enter the national Memory Championship, where he tested his new skills against formidable memory athletes from all over the country in memorizing in only five or fifteen minutes large groups of names and faces, pages of unpublished poetry, sequences of cards, and long strings of random words and numbers. Not only was Scott named that year's national champion, he successfully defended the title in three more consecutive competitions.

Since gaining popularity in the early 1990s, the sport of competitive memory has grown exponentially, with national competitions popping up in every corner of the world. Today, a new generation of master mnemonists are bringing the sport into the twenty-first century, with social media–savvy competitors such as Yänjaa Wintersoul, a Mongolian Swedish triple world-record holder and the first woman to compete on a World Memory Championship–winning team. With her signature fuchsia-dyed hair, Yänjaa is perhaps most famous for a 2017 viral video in which she memorized IKEA's entire furniture catalog (328 pages of roughly five thousand products) in less than a week.

The feats performed by elite competitors such as Scott Hagwood and Yänjaa Wintersoul are even more impressive when you consider that none of these supermemorizers test higher than the average subject in natural memory abilities or even claim to have enhanced mental powers. So how are these mere mortals able to perform such breathtaking feats of memorization? And what do they tell us about the way we all remember?

A clue can be found embedded in the oral traditions that can be traced back over thousands of years. From the *Ramayana* and *Mahabharata* to the *Iliad* and the *Odyssey*, bards and orators committed these classic literary epics to memory through the repetition of patterns in the structure and rhythms of poetry. Similarly, indigenous cultures have preserved and passed down generational knowledge of plants and animals, geography and astrology, genealogy and mythology, through songs, stories, dance, and rituals.

Whether it's today's memory athletes reciting pi to more than one hundred thousand digits, orators of the ancient world entertaining audiences with lengthy tales of heroism, or even a class of kindergartners learning to sing their ABCs, the most effective mnemonics (memory strategies) draw on the fundamental ways in which the human brain has evolved to handle the complexity of the world.

It begins with a process called *chunking*.

MAKE IT CHUNKY

In 1956, George Miller, one of the founders of the then-nascent field of cognitive psychology, wrote a rather peculiar paper. Perhaps indicative of the mores of the time, the paper starts off with a zany, sardonic paragraph that is a far cry from the kind of dry, soulless, prosaic prose that journal editors and cranky reviewers force us to write today:

> My problem is that I have been persecuted by an integer. For seven years this number has followed me around, has intruded in my most private data, and has assaulted me from the pages of our most public journals. . . . There is, to quote a famous senator, a design behind it, some pattern governing its appearances. Either there really is something unusual about the number or else I am suffering from delusions of persecution.

Despite the bizarre tone of the introduction, Miller's paper became a classic because it established a fundamental point about memory that has been validated time and again: the human brain can only keep a limited amount of information in mind at any given time.

Miller used the humorous metaphor of persecution at the hands of an integer to draw attention to his conclusion that we can keep in mind only about seven items. More recent estimates suggest that

Miller was too optimistic and that we can keep up to only three or four pieces of information in mind at once. This limitation of memory helps to explain why, when a website spits out a random series of letters and numbers for a temporary password—say JP672K4LZ—you'll forget it almost instantly if you don't write it down. Professional memory athletes face the same limitation as everyone else, but they get around the problem by exploiting a huge loophole: there is no set definition for what constitutes *one* piece of information. Chunking allows us to compress massive amounts of data into a manageable amount of information that is easily accessible.

Though you might not be consciously aware of it, you're already using chunking in everyday learning and remembering. For example, if you're a U.S. citizen, you have likely committed to memory your nine-digit Social Security number. What makes this sequence of numbers relatively easy to remember is that it's broken down into three memorable chunks—a predictable three-two-four pattern. In the United States, we also remember ten-digit phone numbers (a three-three-four pattern) in a similar way. By grouping those numbers, we reduce the amount of information our brains have to work with by two-thirds. Acronyms (e.g., HOMES for the names of the Great Lakes of North America) and acrostics (phrases such as *Please Excuse My Dear Aunt Sally* for the order of operations in mathematical expressions) follow a similar principle, tying otherwise difficult-to-remember information to simple concepts we can easily grasp. Even that randomly generated, meaningless string of letters and numbers becomes a far more manageable password when you chunk it to JP6-72K-4LZ.

Some of the most compelling research on chunking was done in the 1970s by Herb Simon, a psychologist at Carnegie Mellon University and a pioneer in the nascent field of artificial intelligence. Simon made contributions to many fields, including work in economics that netted him the Nobel Prize in 1978, but to me, his most interesting research was about chess. Simon first became interested in the 1950s in developing computer algorithms to simulate how humans solve problems, and he used chess as the ultimate challenge to solve.

Viewed through the eyes of a novice player, chess can seem intimidating in its complexity. As the game begins, each player has eight pawns, two bishops, two knights, two rooks, one queen, and one king moving around a grid of sixty-four alternating light and dark squares. Looking at the board, beginners might struggle just to keep track of the locations of all their pieces. In contrast, a *grand master*—a title given only to the most elite players of chess—can quickly take in the configuration of pieces on a board, recognizing and responding to familiar patterns and sequences. As a result, a beginner struggles with each move, but the grand master can cut through the clutter and anticipate an entire sequence of moves that has yet to play out.

When Simon studied chess experts, he found that they could look at a set of pieces on a board for only a few seconds and then reproduce the positions of every piece from memory. However, when they were asked to recall the locations of chess pieces placed in random positions that violated the rules of chess, their memory performance cratered to the level of amateurs. These results suggest that chess grand masters do not have extraordinary memory abilities; rather, they rely on the knowledge of predictable patterns and sequences they have accrued across many situations that one would experience in a typical chess game. Like memory athletes, chess grand masters use a combination of skill, training, and experience—aka expertise—to chunk at lightning speed.

In 2004, as my research branched out into multiple areas, I became interested in how expertise changes the way we learn and remember. Most, if not all, of us have some kind of expertise—bird lovers might have the ability to rapidly identify various bird species, whereas car enthusiasts can immediately recognize the year, make, and model of a classic car. At the time, most neuroscientists believed that expertise emerges through changes in sensory areas of the brain. According to this view, avid birders can tell the difference between dozens of different sparrow types because they perceive subtle variations in wing patterning that would seem barely distinguishable to the untrained eye.

As a memory researcher, I had a different intuition. Knowing that the prefrontal cortex helps us to focus on the distinctive aspects of an event, I suspected that expertise changes the way we mobilize the prefrontal cortex. That idea remained dormant until my graduate student Mike Cohen introduced me to a remarkable undergraduate psychology major named Chris Moore. Working together late into the night, Chris and Mike developed a computer program to generate a series of three-dimensional shapes. These shapes looked a bit like alien spaceships, but followed a fundamental structure and logic, in the same way different bird species or makes of cars have certain features that vary and certain features that are, more or less, constant.

Next, Chris and Mike recruited a group of student volunteers, who, over ten days, became "experts" on these alien shapes, learning to identify the common features among the objects as well as to distinguish the differences between them. After the training, we put them in the MRI scanner to see how this training affected their brains. While we recorded their brain activity, they were briefly shown one of the alien shapes and then asked to keep a mental image of it after it disappeared from the screen. About ten seconds later, they were shown another shape and asked if it was the same as the one they had just seen. For someone with no training, this test would have been incredibly difficult, but our volunteers performed nearly perfectly. Like Herb Simon's chess masters, our experts had developed particular ways to extract the most useful information about what they were trying to remember, allowing them to bypass the limitations of memory by leveraging their expertise. However, when we put the shapes upside down and our experts could no longer apply their skills, they found it difficult to tell the alien shapes apart.

As I expected, MRI scans showed that activity in the prefrontal cortex dramatically increased when the students relied on expert skills to keep those alien shapes alive in their memory. What this tells us is that expertise isn't just about seeing patterns, it's about the way we *find* them. For instance, expert birders don't just "see" the difference between a song sparrow and a house sparrow, they use their expertise to home in on the most distinctive features of these birds. As we gain expertise in any topic, we can exploit what we have learned to focus on the most important bits of new information that we need.

Speaking of expertise, I can't help but mention one postscript to this story. While Chris was working in my lab, he was performing poorly in many of his classes. I had no idea because, whenever we talked about memory, he seemed more like a veteran researcher than an undergrad with a mediocre GPA. A few years after graduation, Chris studied neuroscience at Princeton, where he worked on neural network models to simulate learning in the human brain. Instead of focusing on his dissertation research, Chris spent much of his time using his computational expertise to gain insights into the intricacies of baseball, trying to find patterns from baseball statistics to identify great players and winning teams. Chris eventually managed to finish his PhD, and after a brief stint on Wall Street, he put his expertise to use for the Chicago Cubs, where he is now vice president of research and development. His work on predictive analytics earned him the nickname "Moneyball Man," after the complex computational models he created to track and evaluate players' performance helped the Cubs break the seventy-one-year-old "billy goat curse" and win the 2016 World Series.

THE BLUEPRINT

Chunking helps ordinary people, chess experts, and memory athletes alike to compress information into manageable pieces they can work with, but that's only a small part of the story; chunking alone is not enough to explain how any of us manage to remember massive amounts of information without suffering from interference—the competition between memories that accounts for much of our everyday forgetting.

The human brain is not a memorization machine; it's a *thinking* machine. We organize our experiences in ways that allow us to make sense of the world we live in. To handle the complexities of the world, without falling prey to interference, we can exploit one of the brain's most powerful tools for organizing information: the *schema*.

A schema is a kind of mental framework that allows our minds to process, organize, and interpret a great deal of information with minimal effort. The way the human brain uses schemas to construct new memories is not unlike how an architect uses a blueprint to design houses. An architectural blueprint functions as a kind of map of the bare-bones information about the structure (walls, doors, stairs, windows, and so forth) of a building that shows how everything is connected. The abstract nature of a blueprint means it can be reused over and over.

We see this to great effect in the suburban housing tracts that popped up all over the United States in the early 1950s to fulfill the growing demand for cheap housing in the post–World War II era. Drive through any of these planned communities and you will find groups of homes built from the same blueprint. Though the color, window trim, roofing, and so forth may vary significantly, the floor plan and basic structure are the same because architects of this era discovered it was far more efficient and economical (in time, labor, and materials) to use the same blueprint to construct all the homes in a particular area.

In the same way blueprints can be reused to efficiently build new structures, we use and reuse schemas to efficiently form new memories. Though few of us will ever be called upon to memorize an entire furniture catalog, if you've made a few trips to your local IKEA, you probably do have a mental map for the mazelike layout of the store. If your brain stored the memory for the IKEA layout like a photograph, it would be of limited use. But by storing it like a blueprint, you have a mental representation that can be recycled over and over. Even if you go to an IKEA in a new location, the layouts are similar enough that you don't have to form an entirely new mental map to get around. Instead, you can recycle your previous IKEA schema to focus on learning the distinctive features of the new store, enabling you to navigate through the showrooms, then the marketplace and the warehouse, to retrieve your child from the ball pit at Småland.

The concept of a schema doesn't just apply to physical spaces. We all have mental blueprints laying out the sequence of events that are likely to happen in a familiar situation. These "event schemas" provide the structure that allows us to rapidly form memories for a complex event. Suppose you regularly meet a friend for a cup of coffee at a local café. Your brain could record a new, detailed photographic memory for every second of each meeting—waiting in line, ordering a latte at the register, watching the barista pour steamed milk into a perfect rosette. But separately creating an entirely new memory for every detail of each experience would result in potentially hundreds of redundant memories. It's far more efficient to integrate the common features of your café experiences into a single blueprint made up of all the overlapping details. That way, you can focus on learning the stuff that is meaningfully different on each occasion.

In some way or another, every memory athlete, chess expert, birder, and car enthusiast exploits the power of schemas to organize what they need to remember into a framework they can access later. One example is the *method of loci*, an ancient mnemonic device believed to have been invented by the Greek poet Simonides that is more commonly known today as a *memory palace* or *mind palace technique*—used to much acclaim by Sherlock Holmes in the recent BBC adaptation. In this approach, you visualize placing information you want to remember within the layout of a familiar place or route. It can be a palace, a local market, your childhood bedroom—the point is that your schema for that place helps you organize otherwise arbitrary information you can easily access later by taking a mental tour of the location.

Music, much like classic epic poetry, is another wonderful example of how we often exploit organized knowledge to rapidly learn and encode new information. Many songs in blues and rock music follow a typical, repetitive twelve-bar format. Pop and folk music follow a simple verse-chorus-verse structure, and the musical transitions are highly predictable, so if you hear a new song that follows this structure, it's not hard to learn the lyrics or specifics of the tune. Moreover, music is great for memorization. You can easily insert things you want to memorize into musical schemas. Everything I know about the American legislative process comes from listening to "I'm Just a Bill," a *Schoolhouse Rock!* song I heard repeatedly as a child on Saturday-morning television. It's not a stretch to think that a great deal of the music and poetry created by civilizations across the globe has survived over the centuries because of the ease of memorizing and transmitting culturally significant information within a musical structure.

Perhaps the easiest way to use schemas for memorization is as we do in daily life when we are making memories for new events. For instance, if you wanted to memorize the order of a deck of cards, you would not want to memorize each card in isolation. Instead, you can make up a story that relates them to one another (e.g., The king used a jack to replace his flat tire and drove seven miles to the Ace gas station . . .). The effectiveness of such strategies speaks to the intelligent and efficient nature of human remembering, as opposed to the mindless nature of photographic memory.

INTELLIGENT BY DEFAULT

The latest research in neuroscience has revealed a great deal of information about how schemas are implemented in the brain. Ironically, these insights came from the discovery of a network of brain areas that were thought to come online when we are doing nothing.

In most fMRI experiments, subjects are asked to do mundane tasks in which they lie in a scanner and view visual images or words on a screen and make decisions by pressing a button. In the early days, we would interpret these results to suggest that the brain consists of a set of different areas working in isolation, each doing a specific job. As we gained a greater understanding of how the neocortex is organized, however, we could see that is not the case.

Just as human social networks are organized around interconnected webs of family, friends, and work relationships, the neocortex is organized into networks of functionally and anatomically linked regions that communicate with one another as we react and respond to the outside world. As fMRI research progressed, it became increasingly clear that areas in the same brain network tend to activate at the same time. For instance, if I am watching a blank screen and an image of a dog suddenly flashes on the screen, the visual network lights up; if I hear the dog barking, areas in the auditory network light up; and so on. When we are asked to do more attention-demanding tasks, fMRI reveals more activation in various brain networks . . . with one apparent exception.

In 2001, pioneering brain-imaging researcher Marcus Raichle of Washington University observed that a set of neocortical areas consumed the most energy in the brain, yet activity in these areas would go "down" when people were focusing their attention on some arbitrary task, such as pushing a button when they saw an X flash on the screen. Raichle proposed that this network of areas is turned on by default whenever we disengage from the outside world, and he therefore called it the *default mode network* (DMN). By putting a single name on a set of poorly understood areas tucked deep inside the neocortex, Raichle implied that these areas have some kind of shared function.

Neuroscientists tend to have type A personalities, and we take our tasks seriously. Surely nothing useful could come from a network of brain areas that shuts off when people are doing their tasks, right? The DMN is often studied in the context of "mind-wandering" or going "off task"—giving the impression that its main function is to help us goof off or space out.

I was not sure what to make of all this research. Something seemed to be missing. I wasn't satisfied with the idea that evolution would architect a huge chunk of the brain solely dedicated to daydreaming. It was even more puzzling when I learned that brain activity in the hippocampus is tightly linked with what is going on in the DMN. When activity in the DMN goes down, activity in the hippocampus also goes down.

None of this made sense to me until 2011, when I attended a few talks at a memory conference in York, England, and I became clued in to a growing number of fMRI studies that lit up the DMN like a Christmas tree. Although the default mode network appeared to shut down when people did fairly simple tasks (e.g., if I showed you the word *shark* and asked you to say the first verb that came to mind), it lit up when people engaged more complex thought processes, such as calling up autobiographical memories, navigating virtual reality games, or simply trying to make sense of a story or movie. Soon after returning home from York, I teamed up with Maureen Ritchey—then a postdoc in my lab; she is currently a professor at Boston College— to comb through a small mountain of studies that had been done in humans, monkeys, and even rats, and a pattern soon became apparent. We proposed that cell assemblies in the DMN store the schemas we use to understand the world, dissecting the events we experience into pieces so that we can use them in new ways to construct new memories. The hippocampus could, in turn, put the pieces together to store a specific episodic memory.

Although I was excited to test our ideas about the DMN, I did not know where to begin. Almost everything we knew about the neuroscience of human memory came from studies that followed the Ebbinghaus model, in which we asked people to memorize random lists of words or faces. These kinds of tests did not provide much of an opportunity to exploit schemas to their full potential. Fortunately, change was on the horizon. I started to see new results from researchers who were using fMRI to study brain activity as people watched movies or listened to stories. These studies demonstrated that we didn't have to restrict ourselves to capturing microcosms of memory. We could be more ambitious and study memory for the kinds of events that we experience in the real world. I was sufficiently inspired by this work to assemble a team of "superfriends," including Sam Gershman at Harvard, Lucia Melloni at NYU, Ken Norman at Princeton, and Jeff Zacks at Washington University, to build and test a computational model of how the DMN helps us remember real-world events. Remarkably, we convinced the U.S. Office of Naval Research to support this project, and I set out to transform my lab's approach to studying the mechanisms of memory.

We transitioned from studying brain activity as people remembered single words or pictures to more complex experiments in which they would recall the events depicted in a forty-minute movie or story. Our team spent months making movies and writing short stories, and one postdoc, Alex Barnett, even made two animated films (one was a police procedural and the other was somewhere between *Shrek* and *Game of Thrones*). After all this work, we were finally in a position to test hypotheses about how schemas help us understand and form memories about the world.

One of our most interesting studies was done by Zach Reagh, then a postdoc in my lab, and now a professor at Washington University, St. Louis. Virtually all our experiences involve four basic components: people and things interacting in particular places and situations. So, we predicted that schemas for people and things would be kept separate from schemas for places and situations, stored in different parts of the DMN. To test this prediction, Zach became an amateur filmmaker. Using a GoPro camera, he filmed two postdocs in my lab, Alex Barnett and Kamin Kim, each in either a supermarket or a café.

One movie depicted Alex choosing items from the canned-food aisle at the local Safeway, while another portrayed Kamin reading a book and drinking tea at Mishka's, an iconic Davis café. These short clips captured events people could easily understand, so they were perfect for observing if we reuse schemas to understand and remember events. If so, we would expect that areas in the DMN might show the same pattern of activity (i.e., the same memory codes), say, when watching the footage of Alex buying canned beans at the Davis Food Co-op or Kamin buying organic blueberries at Nugget (a fancy local grocery chain). Next, we put people in the MRI scanner and recorded their brain activity as they watched all eight of Zach's films, and again when they recalled the films from memory.

When the experiment was done, we set out to test whether we could read out patterns in fMRI data—memory codes—that could reveal how schemas are reused in different events. We found that the DMN was providing the raw materials needed to understand and remember each movie, but it was not storing context-specific episodic memories. Instead of storing a unique memory code for each movie, the DMN was breaking up each movie into components that were repeatedly reused to understand or remember other movies that shared the same components. Memory codes in one part of the DMN could tell us whether the subject was watching or remembering a movie that took place in a supermarket or a café, whereas memory codes in a different part of the DMN could tell us whether Alex or Kamin was the star in the movie. In contrast to the DMN, the hippocampus, which supports episodic memory by putting together information from all over the brain—did have a separate memory code for each movie. And unlike the DMN, the hippocampus only seemed to store a memory for the beginning and end of each movie (i.e., the event boundaries).

The division of labor in different parts of the DMN suggests that we have different kinds of schemas for different components of an experience. One type of schema can tell us about the context of particular kinds of events, regardless of the specific people involved. For instance, if you are shopping at a supermarket, you know you will have to pay for your groceries, regardless of who's working at the cash register. Another type of schema can tell you about specific people and things. For instance, I have schemas that tell me about who Alex and Kamin are as individuals, regardless of where or when I encounter them. Thanks to the DMN, I can reuse my supermarket schema every time I shop for groceries, and I can reuse my Alex schema every time I see Alex. And thanks to the hippocampus, I can also form different memories for every specific occasion that I run into Alex at the supermarket.

Based on these findings, I have come to believe that forming an episodic memory is a bit like building with LEGOs. You can take apart a LEGO re-creation of a medieval city and sort the bricks and plastic people into different piles. In the same way, the DMN can deconstruct an event, separately processing pieces for *who* and *what* was there and pieces for *where* and *how* the event unfolded. With LEGOs, you can use an instruction sheet to rebuild that medieval scene or use a different set of instructions that show how the same bricks can be combined to reenact a scene from *Star Wars*. Likewise, when it comes to memory, the DMN has pieces that can be reused across many different events. The hippocampus seems to have the instructions for how to put these pieces together to remember a particular event, and hippocampal activation spikes as it communicates with the DMN at event boundaries. So just as you can glance at a set of instructions, build a chunk of a LEGO structure, then look back for direction as you move on to the next chunk, the hippocampus may guide the DMN at those key moments, so that it uses the correct pieces to reconstruct the right memory.

Our research into the DMN has potential importance for our understanding of Alzheimer's and other neurodegenerative diseases. It's now clear that amyloid—a protein implicated in the development of Alzheimer's disease—accumulates in the DMN in about 20 percent of older adults long before any symptoms are apparent. The only way to develop useful treatments for Alzheimer's disease will be to administer drugs to people who are at risk at this "preclinical" stage, because later in the disease, massive cell death occurs in the DMN that can't be reversed. We are currently exploring whether we can use fMRI studies of memory to detect DMN dysfunction early in the disease so that those who are at high risk can receive treatment before experiencing irreversible brain damage.

BACK TO THE FUTURE

If you announced at a dinner party that you could predict the future, you'd likely be met with skepticism. But this claim is actually not so far off from reality. Suppose friends invite you to their teenager's high school graduation; even if you have never attended a commencement ceremony at that school, you would be able to reasonably predict that there will be inspirational speeches and that students dressed in caps and gowns will be handed their diplomas as "Pomp and Circumstance" plays in the background.

Let's reconsider the case of the chess grand master, who has spent countless hours studying and playing out the same patterns over and over across thousands of matches. A grand master has a library of chess game schemas, each containing templates for entire sequences of moves you typically see in a game. Those schemas allow the grand master to remember sequences of moves in past games, to understand what is happening in a game in real time, and to predict likely moves that an opponent could make in the future. By exploiting that expert knowledge, a seemingly complex configuration of pieces can easily be understood as one step in a series of moves that might wipe out a number of pieces and lead to a checkmate.

Professional athletes often resemble chess grand masters in the way that they use their knowledge. In fast-moving team sports such as basketball, football, and soccer, extraordinary physical abilities are not enough. To truly excel, one needs to study the game to build up an arsenal of schemas that can be rapidly deployed as needed. LeBron James—one of the greatest basketball players in NBA history, and holder of the career scoring record—is also known for his ability to accurately recall, in exceptional detail, the sequence of plays in past games. James describes himself as having a photographic memory, but his real strength is what NBA coach (and former Cal basketball legend) Jason Kidd calls "basketball IQ." Like a chess grand master, LeBron draws upon his knowledge of the game to rapidly compress information about complex sequences of actions. In real time, he can cross-reference what he is seeing with his rich mental database of event schemas, so that he can make accurate predictions about the plays that are about to unfold.

As Jason Kidd describes it, LeBron "plays the game that way in the sense of anticipating what's next. And when you have a high basketball IQ, you understand what's going to happen next before anybody else does." LeBron describes his basketball IQ in a similar manner: "It's allowed me to see things before they happen, put guys in position, kind of read my teammates, knowing who is out of rhythm, who is in rhythm, knowing the score, the time, who has it going on the other end, knowing their likes and dislikes and being able to calibrate all that into a game situation."

LeBron leverages the same level of mnemonic intensity when he plays video games with friends. As his longtime friend Brandon Weems put it, "He will know what your game plans were in the past when you've played with him, and he'll pick the opposing team knowing what plays you want to run. . . . You better save your favorite play, too, because he'll remember what you ran before in situations and be ready for it." LeBron's competitive edge comes, in part, from his making optimal use of his memory.

Schemas allow us to see *through* an event, capturing the deeper structures of how everything is connected. In doing so, we can compress memories of hundreds, even thousands, of experiences into a format that enables us to make inferences and predictions about events we haven't yet experienced. Schemas allow us to use knowledge about what *has* happened to get a head start on what *will* happen.

But, as I'll describe in the next chapter, the benefits we gain from such a generative memory system come with a potential cost. If we reuse knowledge across events, what happens when we rely on schemas too much, filling in the blanks of memory in a way that deviates from the reality of what we have actually experienced?

PART 2

———

THE UNSEEN FORCES

. . .

4

JUST MY IMAGINATION

WHY REMEMBERING IS INEXTRICABLY
LINKED WITH IMAGINATION.

. . .

Memory is imagined; it is not real.
Don't be ashamed of its need to create.

—Nick Cave, *The Sick Bag Song*

One of the most expansive memories ever documented belonged to a
Russian newspaper reporter named Solomon Shereshevsky. For much
of his life, he was oblivious of the peculiar nature of his memory.
Then, in his late twenties, the young reporter's habit of never taking
notes during morning staff meetings caught the attention of the editor of his Moscow newspaper. Shereshevsky told the editor he never
wrote anything down because he didn't need to, then repeated ver-
batim the long list of instructions and addresses for that day's assignment. The editor was impressed, but even more interesting to him was
that Shereshevsky seemed to think there was nothing unusual about
this. *Wasn't this how everyone's mind worked?* The editor had never seen
anything like it, so he sent Shereshevsky to have his memory tested.

Shereshevsky then crossed paths with a young researcher, Alexander Luria, at a psychology laboratory at the local university. For
thirty years, Luria, who would go on to become one of the found-
ing fathers of neuropsychology, tested, studied, and meticulously
recorded Shereshevsky's remarkable ability to quickly memorize
made-up words, complex mathematical formulas, even poems and
texts in languages he didn't speak. Even more astonishing than his
ability to recall much of this information with the same accuracy
many years later was that Shereshevsky could remember what Luria

was wearing on the day he had administered a particular memory test. In his classic 1968 monograph, *The Mind of a Mnemonist: A Little Book About a Vast Memory*, Luria wrote, "I simply had to admit that the capacity of his memory had no distinct limits."

Luria linked Shereshevsky's remarkable capabilities to an extremely rare condition called synesthesia—meaning that every stimulus, regardless of which sense it came through, triggered every other sense. Shereshevsky could taste words, see music, and smell colors—even the sounds of words could impact his perception. He described asking an ice cream vendor what flavors she had. Something about her tone of voice made him see a stream of black cinders pouring from her mouth as she spoke, "Fruit ice cream"—which promptly ruined his appetite. The connection between the worlds he created in his mind and the world he lived in was so visceral that Shereshevsky could elevate his heart rate by simply imagining he was running after a train. He could raise the temperature of one hand and lower the other by picturing one hand on a stove and the other resting on a block of ice.

The distinctiveness of Shereshevsky's sensory world extended to his imagination, giving him the ability to form distinctive memories that were resistant to interference. *New Yorker* writer Reed Johnson, who spent years researching Shereshevsky, described how he was able to attach any memoranda, no matter how bland, to stories he conjured in his imagination, which he could follow like a trail of bread crumbs to find his way back to that information later on:

> The strength and durability of his memories seemed to be tied up in his ability to create elaborate multisensory mental representations and insert them in imagined story scenes or places; the more vivid this imagery and story, the more deeply rooted it would become in his memory.

In his later years, when Shereshevsky began performing his incredible memory feats for a paid audience, he amplified this ability with a technique familiar to modern memory athletes such as Scott Hagwood and Yänjaa Wintersoul. Though it appeared to have been self-discovered rather than learned, Shereshevsky used a memory device similar to the method of loci. When he wanted to remember a sequence of words or numbers, he would visualize them as characters within the familiar schema of, say, a street in Moscow and take a "mental walk" through the vast worlds of his interior landscapes.

Although he is often discussed as an example of someone with an extraordinary memory, the key to Solomon Shereshevsky's mnemonic capabilities was his vibrant imagination. Much of Luria's decades-long study reveals a fundamental truth about the connection between memory and imagination, one that lies at the center of how we all remember. In this chapter, we'll explore how the peculiar way in which we form memories can lead us to stray far from reality, yet gives us the fuel to imagine a world with endless possibilities.

WHAT *CAN* HAPPEN

The simplest way to see the machinery of episodic memory at work is to scan people's brains while they describe an event from their lives. For instance, if you showed me the word *photograph* while I was lying in an MRI scanner and asked me to use that word to help me recount an event from my life, I might pull up the memory of my first live rock concert. At fourteen, I was obsessed with the album *Pyromania* by British heavy metal band Def Leppard. If you examined my brain activity while I recalled seeing Steve Clark play the signature riff during the band's performance of "Photograph," you'd see activation in the hippocampus, as I pulled up the contextual information that mentally transported me back to 1985, and in the DMN, as I brought up knowledge about concerts that enabled me to elaborate on how the event had unfolded.

Now, let's try something a little different. Suppose you were lying in an MRI scanner and I showed you words such as *pasta* or *skydiving* and asked you to use those words to *imagine* something that hasn't happened, or even something that would be unlikely ever to happen. You might conjure up a mental image of cooking spaghetti with Motown legend Marvin Gaye or jumping out of a plane with pioneering physicist Marie Curie. In 2007, three research labs published experiments along these lines, and here's the twist: the brain activity changes that occur when people imagine these kinds of scenarios are remarkably similar to those that occur when people recall events that they actually experienced.

This odd parallel between imagination and memory came as a surprise to many in the scientific community, and it captured the attention of the media—*Science* magazine declared it one of the top ten breakthroughs of the year —but it did not come out of the blue. It was anticipated nearly a century earlier by English psychologist Sir Frederic Bartlett, whose work would become the foundation for the idea that we use mental frameworks (i.e., schemas) to organize and process the world around us.

Bartlett began his research on human memory as part of his dissertation at the University of Cambridge in 1913. After receiving his PhD, he focused not on memory but on cultural anthropology and then on applications of psychology for the military. Fortunately, Bartlett eventually circled back to the topic of memory and, in 1932, published his most important work, *Remembering: A Study in Experimental and Social Psychology*.

Bartlett's book was a dramatic departure from the tradition of memory research established by Hermann Ebbinghaus back in 1885. Ebbinghaus quantified memories for strange, meaningless information under strictly controlled conditions. In contrast, Bartlett drew on his experiences in applied psychology and anthropology, observing and describing how we use memory in our everyday lives. Put more succinctly, Bartlett was interested in understanding *how* we remember, rather than simply quantifying *how much*.

In his most famous experiment, Bartlett introduced a group of volunteers at the University of Cambridge to a Native American folktale called "War of the Ghosts"—specifically chosen because the cultural context was entirely foreign to these British students. Bartlett's subjects could recall the gist of the story, but they made some characteristic errors. It was not simply a case of failing to remember some details, but rather that they adapted the details to match up with their own cultural expectations and norms. Words such as *canoe* and *paddle* were replaced by *boat* and *oar; seal hunting* became *fishing*.

Poring over these results, Bartlett observed that, although people do recall some details from the past, their recollections are approximate at best. He concluded, "Remembering is not the re-excitation of innumerable fixed, lifeless and fragmentary traces. It is an imaginative reconstruction." We do not simply replay a past event, but use a small amount of context and retrieved information as a starting point to imagine how the past could have been. We put together a story on the fly, based on our personal and cultural experiences, and tack on those retrieved details to flesh out the story. Bartlett's insight is key to understanding why the brain's machinery for imagination and its machinery for memory aren't completely independent—they are both based on pulling up knowledge about what *can* happen, though not necessarily what *did* happen.

FABLES OF THE RECONSTRUCTION

The reconstructive nature of memory means that our recollections can sometimes take on a life of their own. Consider the case of former NBC news anchor Brian Williams. Speaking at a New York Rangers hockey game in 2015, Williams recounted how, in 2003, he and his crew were in a helicopter in Iraq that was forced down by a rocket-propelled grenade. His account was promptly debunked by several veterans who had been there. Williams had never encountered enemy fire during his visit to Iraq—though he certainly found himself at the center of a firestorm of controversy after what seemed like a blatant lie.

In reality, Williams and his crew were flying about an hour behind a convoy of three military helicopters, one of which was hit by a rocket-propelled grenade; all three were to make an emergency landing in the desert. Williams's helicopter eventually caught up to the other three, and a sandstorm stranded them in the desert together for several days. Although elements of Williams's 2015 account overlapped with what really happened, the story he told twelve years later didn't belong to him but to the soldiers in the helicopter that had gone down. Williams apologized, chalking it up to the "fog of memory," but the damage was done. Widely suspected of lying to enhance his reputation, Williams's integrity as a journalist was tarnished. He was suspended without pay for six months and ultimately stepped down from his role at *NBC Nightly News*.

I can't say whether Brian Williams intentionally embellished his story, but if we give him the benefit of the doubt, it appears that he recollected many of the right pieces, yet reconstructed a narrative of the event that was fundamentally wrong. His dramatic account was an imaginative reconstruction gone awry.

Most of the time, our memories do not stray as far off the mark as Williams's Iraq anecdote, but ample scientific evidence suggests that we can confidently remember stuff that didn't happen. In 1995 Henry Roediger III and Kathleen McDermott of Washington University, St. Louis, demonstrated this phenomenon in an experiment that is now taught in almost every introduction to psychology class. They had volunteers study lists of words such as this:

FEAR
TEMPER
HATRED
FURY
HAPPY
ENRAGE
EMOTION
RAGE
HATE
MEAN
IRE
MAD
WRATH
CALM
FIGHT

Now, indulge me for a moment. Without looking back at the list, what words do you recall seeing? Do you remember reading the words *fear* and *wrath*? What about the word *anger*? If you remembered seeing the latter, you would be mistaken—but you wouldn't be the only one. In fact, the people who participated in Roediger and McDermott's study were just as likely to remember seeing *anger* as they were to remember words that they had actually studied, such as *fear* and *wrath*.

This finding is often used by scientists to describe how people are susceptible to "false memories," and the potential implications are staggering. It's easy to read about studies like this and go down a rabbit hole of self-doubt, wondering whether our most cherished memories might be entirely false. But that's not quite the right way to think about it.

As former Sex Pistols / Public Image Ltd singer John Lydon put it, "I don't believe in false memories, like I don't believe in false songs." In the real world, memory can't easily be reduced to simple, black-and-white dichotomies such as strong or weak, true or false. When people with a healthy brain get a little feeling of mental time travel and say, "I definitely remember that," they are probably pulling up *something* from their past. However, even when the elements of what we remember are true, the entirety of the story can be false. Roediger and McDermott's experiment was explicitly designed to encourage people to think about anger, or maybe even think about a moment in their lives when they felt angry. So, when their subjects (or you) recalled that the word *anger* was among those on the list, it was, in some sense, a real memory, but one that was reconstructed incorrectly. Perhaps this is also what happened to Brian Williams.

Not everyone is equally susceptible to generating false memories. Individuals with amnesia generally don't fall for the trick of recalling *anger* because they aren't able to recall enough information to make a reconstructive error. Some research suggests that people with autism or other neurodevelopmental disorders are also more resistant to false memories because they sometimes remember events in a concrete manner, focusing on the details at the expense of the meaning. These studies indicate that complete resistance to false memories might come at the expense of the capability to generate meaningful reconstructions of the past.

All this work points to a deeper realization about memory. When we talk about memories as being "true" or "false," we are fundamentally mischaracterizing how memory works. I like to think of memory as less like a photograph and more like a painting. Most paintings typically include some mixture of details that are faithful to the subject, details that are distorted or embellished, and inferences and interpretations that are neither absolutely true nor entirely false, but rather a reflection of the artist's perspective. The same is true of memory.

Memories, we will see, are neither false nor true—they are constructed in the moment, reflecting both fragments of what actually transpired in the past and the biases, motivations, and cues that we have around us in the present.

FILL IN THE BLANKS

We like to think of memories as though they are tangible. As if somewhere in the vault of our brain there is a complete record of every event we have experienced. As I mentioned in chapter 2, the hippocampus does appear to record a kind of index, enabling us to find the various cell assemblies across the brain that were active during a past experience, so we can reboot our brain to the state it was in during that event. Neuroscientists often focus on this aspect of memory, casting remembering as the turning on of the exact collection of neurons that were active during an event. Frederic Bartlett, with his conviction that memory is an act of "imaginative reconstruction," took a radically different view. He explicitly eschewed the idea of memory traces, arguing instead that memories are born in the moment of reconstruction. In other words, it doesn't make sense to talk about a single memory for an event when innumerable possible memories can be constructed to describe the same experience.

I definitely believe that the hippocampus enables us to get into a past mindset and pull up some details from a past event. But I also agree with Bartlett's contention that, once we get back to the past, we don't simply replay things as they happened. If that were the case, while recalling a ten-minute phone conversation, you would spend ten minutes reliving everything you experienced during that conversation. That's not what happens. Instead, we typically compress that experience into a shorter narrative that captures its gist. So, the hippocampus might get us back to *some* of the cell assemblies that were active during *some* moments in that conversation, but we still need to use schemas in the default network to make sense of what we are pulling up. This reconstruction is prone to error, however, because schemas capture what typically happens, not what did happen.

When we remember, we're like detectives, trying to solve a mystery by piecing together a narrative from a limited set of clues. A detective can build a case based on an understanding of the killer's motive, which can be helpful, but it can also lead to biases. In a similar vein, when we remember events, motive can exert a powerful explanatory role, helping us make sense of what occurred. It infuses action with meaning, allowing us to gather up threads of information and weave them into a memorable narrative. But assumptions about people's motivations can also fuel our imagination, leading us to fill in the blanks about events in ways that distort our narratives of what happened.

Another factor that can bias our memories is that our own goals and motivations affect how we reconstruct an event. I'm often asked, "How is it that two people can experience the same event together and yet recall it so differently?" To quote Ben Kenobi from *Star Wars*, "Many of the truths we cling to depend greatly on our own point of view." People's different goals, emotions, and beliefs lead them to interpret an event from particular perspectives, and those perspectives will also shape how they reconstruct that event later on. For example, two people might watch the same World Cup match together yet remember it quite differently. In a 2017 study, rival fans of two German soccer teams watched the same Champions League final game, yet their memories were so biased that each side recalled their own team gaining more ball possessions and the opposite team gaining fewer. Fortunately, soccer fans notwithstanding, we are not doomed to live in siloed realities. We can break through the walls of our perceptions and find common ground when we assume the other person's perspective while reconstructing the event. Shifting perspective can also help you recall pieces of information that you couldn't pull up before.

Distortions in memory can also be driven by external factors, and it only takes a little nudge to influence our reconstructions. In the early 1970s, Elizabeth Loftus, then an assistant professor at the University of Washington, became interested in how witnesses recall events in the courtroom, and whether their testimony can be biased by leading questions from an attorney. To answer this question, she had a group of volunteers view short films of seven traffic collisions and then asked them to estimate the speed of the cars. Without access to the actual speedometer readings, they could only make guesses based on what they remembered.

Loftus found that it was remarkably easy to bias those guesses. One group of volunteers was asked to estimate how fast the cars were going when they "contacted" each other. Their average estimate was about thirty-one miles per hour. Another group was asked how fast the cars were going when they "smashed into" each other. The average estimate for this group was about forty-one miles per hour. By simply changing the verb she used in her question, Loftus increased the estimated speed by about 33 percent.

Loftus's study highlights the ways eyewitness testimony can be corrupted in the courtroom, but the ramifications are even greater. If even subtle hints and cues can affect the stories we construct about our past experiences, then our memories are reflections of both the past and the present. Our thoughts and motivations at the moment of recollection can influence our memories of the past.

Given the evidence that we reexperience the past by imagining it, and that the process is vulnerable to influence, we are faced with a difficult question: How can we know the difference between fact and fantasy? How do our brains navigate the borderless landscape of our imagination while simultaneously keeping us anchored in reality?

IS THIS REAL LIFE, OR JUST FANTASY?

As anyone who has tried to meditate for more than a few minutes can tell you, our minds are constantly churning with what-ifs. We conjure up scenarios for what could happen in the future, and we wonder what our present would be like if past events had turned out differently. My former mentor and collaborator Marcia Johnson, director of the Memory and Cognition Lab at Yale University, was among the first in the field of psychology to realize that this creates a huge problem. All the scenarios we imagine leave us with memories of events we have never experienced, and our memories don't come with labels certifying them as imagined or real. So, it's not a question of whether we get confused between memory and imagination, but rather, what factors keep us from making such mistakes all the time?

Over the course of her career, Marcia delved into the processes by which we distinguish between internally generated information (thoughts, feelings, imagination) and the information we take in from the outside world. Her work has shown that we can keep some of the inaccuracies of memory in check by consciously applying some critical thinking to our imaginative reconstructions, a process she called *reality monitoring*.

Two factors can help us when it comes to reality monitoring. One is related to differences in the features generated by real versus imagined events. On average, imagined events are more focused on our thoughts and feelings and are less detailed and vivid than events that we actually experienced. As a result, we are more likely to believe real events than imagined elements as we are remembering.

Here's an example of how reality monitoring might play out in real life. If I asked you to think of your last visit to the doctor, you might call to mind your irritation at the stack of medical forms you were asked to fill out before being called into the examination room, your anxiety as you waited for your doctor to come in with the results of your lab tests, or your reluctance to confess that you haven't been taking a prescribed medicine. Now, you might have had those thoughts and feelings even if you just imagined going to the doctor's office. On the other hand, perhaps as the memory unfolds, you hear the annoyingly cheerful Muzak in the waiting room, the antiseptic smell permeating the examination room as you changed into a scratchy paper gown, or the cold of the stethoscope on your chest. The more sensory details that come to mind when you remember an event, the more likely it is that it really happened, because on average what we imagine is not as detailed as what we have experienced.

The other factor that can help with reality monitoring has to do with the way we go about evaluating our mental experiences. A particularly vivid memory constructed with imagined elements (especially elements that fit our schemas or motives) can lead us astray, but you can stay on track if you're willing to put in the work to gauge the accuracy of your memories. Take a moment to ask yourself, "Am I pulling up the actual sights and sounds from a past event or just thoughts and emotions I could have experienced when I anticipated an event or thought about it after the fact?" We can counteract the fallibility of memory by considering not only the quality of the details that seem to put us back in a specific place and time but also the likelihood that those details could have been constructed to create an alternative reality. As with all critical thinking, it helps to remain skeptical until presented with further evidence.

Anything that affects the prefrontal cortex is going to affect your ability to remember with intention, and that also applies to reality monitoring. Marcia Johnson's ideas inspired my dissertation research in graduate school, which identified the prefrontal cortex as a key region in using intention to monitor the accuracy of our memories, and a few years later Marcia and I collaborated on an fMRI study showing that monitoring the details of our memories engages the most evolutionarily advanced areas of the human prefrontal cortex. Our findings coincided with similar results from a number of other labs. Later, my friend and collaborator Professor Jon Simons at the University of Cambridge discovered that people who have more gray matter in this area are better at reality monitoring.

At the opposite end of the spectrum, some people who have extensive damage to the prefrontal cortex can confidently recall things that never happened, a phenomenon called *confabulation*. Neurologist and author Jules Montague described the case of a young seamstress from Dublin, Ireland, named Maggie who was convinced she had visited Madonna's house the week before and advised her on what outfits to wear on tour. Although Maggie had never met the pop singer, she was not psychotic nor was she lying; she had encephalitis, swelling of the brain, which was interfering with her ability to monitor the sources of information that popped into her head.

We all are guilty of minor confabulations. When we're tired or stressed-out, or when our attention is divided by multitasking, reality monitoring goes out the window. As we get older, prefrontal function gets worse, and we find it harder to tell the difference between imagination and experience. This happens even to memory researchers. Every morning I am greeted with a flood of emails in my in-box. Often, I will see a message that requires me to do some task, and I make a mental note to get to it later when I have time to respond. After going through all the messages in my in-box, however, I will frequently confabulate that I responded to the email, only to be surprised a week later when I get an exasperated follow-up message from the same sender with *URGENT* in the subject line.

Reality monitoring can also be challenging for people with vivid imaginations. If we visualize something in extraordinary detail, we can drive the sensory areas of the brain in a way that's similar to what happens when seeing something in the outside world. Accordingly, it can be extraordinarily difficult to tell the difference between experienced events and events that were vividly imagined.

Solomon Shereshevsky, whose remarkable memory was inextricably tied to his extraordinarily vivid imagination, struggled to walk this line. He noted "no great difference between the things I imagine and what exists in reality." After decades studying Shereshevsky, Alexander Luria concluded, "One would be hard put to say which was more real for him: the world of imagination in which he lived, or the world of reality in which he was but a temporary guest." Ultimately, the same magical thinking that allowed Shereshevsky to control his heart rate or body temperature, even to imagine away pain or illness, alienated him from the world around him. Although for a time he had some success performing as a traveling mnemonist, his mental powers never led to the greatness he had imagined for himself. Eventually, he married and had a son and, in his later years, made a living as a taxi driver in Moscow, but his reality never quite measured up to the vast worlds he had created in his mind. He is said to have died in 1958 from complications of alcoholism.

SPARK OF CREATION

Close your eyes and imagine yourself on a white sandy beach by a beautiful tropical bay. Perhaps you visualized lying in a hammock under the shade of a palm tree, sipping a mai tai and watching waves of turquoise water gently lapping against the shore. If you're like me, you probably got so caught up in this tropical fantasy that, for a moment, you lost track of the outside world. Now, consider how one individual with amnesia imagined the same scenario: "Really all I can see is the color of the blue sky and the white sand." Even with repeated prompts, patients in this study struggled to come up with anything that would conjure up a coherent and detailed image of the scene.

Frederic Bartlett not only believed that remembering is an imaginative reconstruction, he also argued that imagination is a product of memory. In a 1928 paper that was far ahead of its time, Bartlett speculated that creative works are constructed by essentially doing the opposite of reality monitoring—that is, by pulling up fragments of memories and then, much like Maggie's Madonna confabulation or Shereshevsky's mental walks, assembling those bits and pieces into a cohesive product of imagination. Bartlett's paper consisted mostly of descriptive observations and speculation, but there is more compelling data for the idea that memories serve as the raw materials for creative inspiration.

Results from fMRI studies and work done in people with amnesia suggest that the mental processes we use to vividly imagine ourselves in a situation, such as sipping a cocktail in a hammock on a tropical beach, are similar to those that we draw upon when recollecting the past. We pull up details from a few different events in our past (via the hippocampus) and rely on information from schemas (via the default network) to assemble a story about how it all could fit together.

The takeaway from this research is not that we generate distorted, inaccurate memories because of our wonky brains. Instead, research by neuroscientists Daniel Schacter and Donna Addis suggests that the hippocampus and the DMN might function at the crossroads between memory and imagination by allowing us to extract the ingredients from past experiences and recombine them into new creations. For instance, fMRI studies have linked activity in the hippocampus and DMN to performance on laboratory tasks designed to elicit creative thinking, and conversely, dysfunction in these areas seems to impair performance on such tasks. And, consistent with Bartlett's ideas about imagination and memory reconstruction, people who show higher performance on tests of creative thinking also are more susceptible to so-called false memories. There seems to be an inextricable link between reconstruction and creation.

These findings have given me, as a songwriter and musician, a new perspective on creativity. I sometimes feel stuck in a creative rut, adrift waiting for a new idea to pop into my head. But a new work of art, music, or literature doesn't arise out of thin air—it emerges from the process of integrating elements from different past experiences. As *Steal Like an Artist* author Austin Kleon put it, "The idea that the artist should sit down and create something 'new' is a paralyzing delusion. We can only create a collage of our influences, our memories—filtered through our imagination."

The schemas that form the backbone of our episodic memories also serve as raw materials for storytelling. In *The Hero with a Thousand Faces,* literature professor Joseph Campbell proposed that every culture has a story that follows a universal narrative arc, in which an ordinary person (almost always an orphan or outcast) is called to leave familiar circumstances, navigate an unfamiliar world, do something extraordinary, and eventually save the day. Virtually every blockbuster film of the late twentieth century, from *Star Wars* and *Spider-Man* to *The Lion King* and *The Lord of the Rings,* follows this narrative blueprint. These movies layer enough content variation onto a familiar structure to allow audiences to feel the excitement of experiencing something new within the comfort of a predictable story arc.

The link between imagination and memory also shows up in the field of generative artificial intelligence, where companies are creating increasingly sophisticated programs, trained with massive numbers of examples of a particular genre, that can generate new outputs that seem to capture the essence of what they were trained with. One of my favorite examples is a project called Relentless Doppelganger, a nonstop livestream of AI-generated death metal that is capable of producing songs that could have been recorded by any of a hundred black metal bands from Scandinavia. Another program, Stable Diffusion, uses AI to respond to user prompts such as "monkey eating a banana split" to create new works of visual art. Other AI-based products, such as controversial chatbot ChatGPT, can make entire poems or stories based on simple user inputs. The results can be amazing, but it's not necessarily surprising. AI art is not about generating something new, it's about taking elements from preexisting human art and recombining them (based on human direction and curation).

Should artists, authors, and musicians be worried about the rise of the machines? No. If anything, AI art is just a reflection of the cultural milieu that spawns human-generated artwork. As I described in chapter 2, the neural networks powering modern generative AI are inspired by neocortical networks in the human brain that are optimized to learn the general structure of what we have experienced. So, it makes sense that a computer program with no understanding of the artistic process can capture the essence of human-generated art that fits within an established genre.

Unlike a computer program that finds common elements among examples of numerous artworks from a particular genre, more innovative artists draw from an eclectic range of influences, often surrounding themselves with people from different backgrounds and exposing themselves to a broad range of creative works from different genres, eras, and cultures. Innovative artists discover connections between movements and ideas that would traditionally be labeled as distinct (by conventional humans and generative AI). We see this tapestry of eclectic influence reflected in creative genius across all forms of artistic expression. Pablo Picasso's cubist paintings were preceded by a period during which he was profoundly influenced by traditional African sculpture and masks. Japanese auteur Akira Kurosawa, who spent his formative years studying Western painting and cinema, used light and shadow to compose shots in his films in a way that is reminiscent of the brushstrokes of expressionist painters. Members of the rap group Wu-Tang Clan grew up on Staten Island in New York City, but their groundbreaking music and lyrics are a mash-up of cultural influences ranging from comic books and science fiction, to Chinese martial arts films, to the philosophy of the Nation of Islam. By exposing ourselves to a diverse range of people and ideas, we can discover new connections and recombine our experiences into new artistic constructions that transcend the sum of their parts.

What makes great art both singular and universal is that it offers a version of reality marked by the idiosyncrasies of its creator, rather than a perfect recording. The same can be said of memory, and for much the same reason; our memories reflect both what we experienced and our interpretations of what happened.

The science of memory is a still-unfolding frontier. Over the last century we have broadened our understanding of how the neural networks that have evolved to remember the past intersect with those allowing us to imagine the future. We now know that it is in that space where memory meets imagination that we interpret reality and create our greatest treasures.

5

MORE THAN A FEELING

WHY OUR MEMORIES ARE DIFFERENT FROM THE FEELINGS WE HAVE ABOUT THEM.

. . .

I think the highest and lowest points are
the important ones. Anything else
is just . . . in between.

—Jim Morrison

We tend to disproportionately remember the highs and lows from the past. And we don't just remember *what* happened; our memories of extreme experiences are often accompanied by raw, visceral feelings that can sometimes be overwhelming. The physical sensations and emotions that bubble up when we revisit emotionally significant events imbue those memories with a sense of immediacy and urgency, giving us the perception that *what* we remember is inextricably linked with the *feelings* that accompany them. I thought that was the case, too, until I started working in a clinic with people struggling in the present with emotions they carried from the past.

In 1998, as part of my work toward a PhD in clinical psychology, I spent a year as a psychology intern at Chicago's Westside Veterans Administration hospital. Like the first year of a medical residency, a clinical psychology internship transitions you from spending most of your time on research and classes to working full-time in a real clinical environment. After my clinical training in the posh suburb of Evanston, Illinois, working at the VA seemed like an entirely different world. It opened my eyes to the powerful influence emotions exert on memory.

Most of my patients were male veterans in their midfifties who had survived the horrors of war in Vietnam and were now living in neighborhoods in Chicago that were being ravaged by gang violence and police brutality. Stories of pain and loss were etched on their faces, though they weren't always able to articulate the source of their trauma. This was especially true of L.C., a relatively young female veteran of the first Gulf War.

I first met L.C. while I was doing intake in the VA's Day Hospital, an intensive outpatient program for veterans in crisis. She had steel-blue eyes, a weathered face, and a square jaw set in an expression of stubborn independence. I asked her questions to get a sense of her emotional state, and her terse responses gave me the impression that she was not comfortable acknowledging any kind of vulnerability, let alone reaching out to another for help. That she was speaking to me in the hospital indicated that she had run out of options. She said that she felt constantly on edge. She was hardly sleeping, and when she did sleep, she was awakened by terrible nightmares. She had been living on a razor's edge and was now plagued by suicidal thoughts. I knew she hadn't seen active combat, and she wouldn't say much about her time in the war, so my challenge was to get her to trust me enough to share more about her experiences.

With time and patience, I did gain L.C.'s trust. Eventually, she revealed that she had served on a mortuary unit during her tour of duty in Iraq. Her job was to prepare the bodies of dead soldiers to be sent back home to their families in the United States. During our sessions, L.C. vividly described the smell of death in the air, and the horrors of dealing with bodies that had been ripped to pieces by improvised explosive devices (IEDs). After her tour was over and she returned home, she was constantly anxious, haunted by recurring memories of the corpses of fallen soldiers. Because L.C. had never seen combat, it hadn't occurred to her, or to the VA, that she might be suffering from post-traumatic stress disorder (PTSD). But she had clearly been subjected to extreme trauma by her experiences in the war. The tremendous intensity of the emotions L.C. contended with daily during her assignment created vivid memories that plagued her constantly.

I wish I could say L.C.'s experience was unique, but most of my patients at the VA had some form of PTSD. Although many of us get frustrated by our inability to remember the past, those with PTSD suffer the burden of remembering it *too well*. They reexperience their trauma repeatedly through flashbacks and nightmares. In addition to combat survivors, PTSD is often experienced by survivors of child abuse, sexual assault, car crashes, and natural disasters. Many New York firefighters on the scene of the World Trade Center terrorist attack on 9/11 and first responders in emergency rooms throughout the world during the COVID-19 pandemic experienced PTSD.

PTSD has both individual and societal consequences, resulting in high rates of alcoholism and drug abuse, unemployment, and homelessness among sufferers as they struggle to cope with debilitating trauma. L.C.'s case is an extreme example, but many of us are affected by painful memories we carry around that can affect how we feel, think, and act in the present.

The power of emotion's influence on memory isn't always negative. Emotional arousal works the other way, too. Think of when you met the love of your life or the birth of your child, and you can feel the emotional intensity that infuses our most memorable experiences. But how and why do our emotions color the way we remember our past, and more important, how do emotional experiences in our past affect us in the here and now? As we will see, the brain mechanisms for remembering what happened are different from those responsible for the feelings that bubble up when we remember, and that distinction has significant implications for our views of the past and the decisions we make in the future.

THE HEAT OF THE MOMENT

Why do our most emotionally intense experiences—times when we were livid with anger, petrified with fear, or in shock from witnessing something terrible—seem to be indelible in memory? The answer to that question is fundamental to the very reason we evolved the capability to remember: our memories hold the key to our survival.

As we've learned, the brain is constantly prioritizing what it thinks is important and letting us forget what isn't. So, it makes sense that we tend to remember the events associated with intense emotions, but that's only part of the story. Emotions, the conscious feelings we experience based on myriad combinations of internal and situational factors, are central to the human experience, but in and of themselves they aren't necessarily important for our survival—feeling a little guilty or embarrassed isn't going to put food on the table or a roof over your head. Rather, the critical influence of emotional experiences on memory has to do with what neuroscientist Joe LeDoux calls *survival circuits.*

Our emotions, as well as the actions and choices they influence, are shaped by basic survival circuits in the brain that motivate us to avoid threats, find sustenance, and reproduce. When these circuits go into overdrive, we tend to experience intense emotions, such as elation, lust, panic, anxiety, or disgust. It makes perfect sense that these are experiences we remember most vividly. Events that intensely activate our survival circuits are worth remembering because they usually provide valuable information that we can use in the future to stay safe, thrive, and reproduce. We might not have survived as a species if our cave-dwelling ancestors hadn't found encounters with saber-toothed tigers to be particularly memorable.

When a survival circuit in the brain gets revved up, say by the terror of a face-to-face encounter with a predator, or the joy of holding your child in your arms for the first time, your brain gets flooded with *neuromodulators.* Neuromodulators are chemicals that influence the functioning of neurons, but they don't simply increase or decrease neural activity. Neuromodulators have more complex effects that fundamentally change how neurons process information. Some neuromodulators are like the hot sauce on your tacos—they change the flavor, add heat, and cause us to sit up and pay attention. Neuromodulators also promote *plasticity,* meaning they enable significant, long-lasting changes in the connections between neurons in the cell assemblies that are activated when we learn something new.

Noradrenaline (also known as norepinephrine) is one well-studied neuromodulator that influences how we learn and remember. You've probably heard of the fight-or-flight response. When we experience threats, the adrenal glands mobilize us into action by pumping out adrenaline, which raises our heart rate, blood pressure, and breathing rate. Noradrenaline, in turn, is released all over the brain. Adrenaline and noradrenaline are the chemical costars of the fight-or-flight response, contributing to the feelings of arousal and immediacy you might experience if you go bungee jumping or get into a shouting match with a driver who cut you off.

Psychologist Mara Mather has shown that emotional arousal ratchets up the stakes for attention, making us better at perceiving the salient things that are important or stand out in some way. So, emotional arousal influences the outcomes of those "neural elections" that determine what we will perceive, funneling resources to the strongest candidates.

Because emotional arousal constrains what we will pay attention to, we can expect that it should change *what* we will remember, rather than simply *how much*. For instance, if you're getting mugged at gunpoint, your attention will be captured by the weapon pointed at you, possibly at the expense of the shoes the mugger is wearing. Just as increasing the contrast in a photograph makes some information stand out and pushes other information into the background, noradrenaline jacks up the contrast of our memories, highlighting the significant details.

The effects of noradrenaline keep going even after an emotionally intense event has passed. Over a few hours, noradrenaline triggers a cascade of events in the cell assemblies that were active during the event, turning on genes that manufacture proteins that ultimately tighten up the connections between those neurons so the memory remains robust over time. If you witness a horrible car crash outside your local grocery store, the release of noradrenaline promotes changes in the connections between cells in your brain so your memory for that particular trip would be more likely to stick around than if you had walked into the store without incident. This is a key reason it's so easy to forget more mundane experiences yet so hard to let go of a traumatic memory: our brains are designed to hold on to the events that revved us up, ostensibly because remembering those events has survival value.

FEAR AND LOATHING IN THE AMYGDALA

Emotional arousal does more than simply affect the memorability of an event. When we recall a traumatic experience, we don't just remember what happened, we reexperience the sensations vividly and with immediacy. It feels as if the memory is stored in our bodies as well as our minds. To survivors of combat situations or sexual assault, the physical experience of reliving those traumatic events can become debilitating, effectively retraumatizing them each time the event is recalled.

This visceral experience of remembering emotional events hinges on a key part of the brain: the *amygdala*. This almond-shaped area (hence the name, which is Latin for "almond") sits just inside the temporal lobes, right in front of the hippocampus, and it is a core part of the brain's circuitry that responds to threats. The amygdala has a close line of communication with the different neuromodulatory systems in the brain as well as the peripheral glands that bring about the fight-or-flight response. When we recall a time our survival circuits were called into action, whether it was a harrowing car crash or the adrenaline rush of zip-lining through a forest, the amygdala and hippocampus work as a team. As the hippocampus forms memories that capture the context at that moment, the amygdala is connecting those memories with the survival circuits that generate the raw sensations. Later on, as the hippocampus helps us mentally time travel back to those events, the amygdala brings us back to the heat of the moment, making us feel as if we are vividly reexperiencing the event.

The division of labor between the hippocampus and the amygdala suggests that there is an important distinction between *what* we remember and how we *feel* when we remember. For instance, people with amygdala damage due to a rare disease called Urbach-Wiethe syndrome can remember past events, but unlike individuals with a functioning amygdala, they don't seem to respond any differently to a gruesome photo of surgery on a car crash victim than they would to a mundane picture of a mother and child in a car. Conversely, people with damage to the hippocampus can have no memory of getting an electric shock but still experience an unconscious threat response when reminded of the event. In the healthy brain, the hippocampus and the amygdala work together, so we can recall what happened and also reexperience how we felt when it happened.

Our understanding of how the amygdala and hippocampus work

gives us insights into how we can understand and process memories for emotionally painful events. That vivid jolt we get upon remembering an emotional experience can trick us into thinking that it is fundamentally a part of our memory of what happened, but that is not the case. In my work with L.C. and others, recovery was not about forgetting the past but rather about managing the intense emotions brought about by the past.

The Day Hospital team spent several weeks learning more about L.C.'s past and working through how her past traumas were affecting her in the present. Individuals who have endured extreme trauma, as L.C. had, tend to disconnect from their thoughts and feelings in the present day, a hallmark of PTSD known as *dissociation*. As part of her treatment, L.C. worked on grounding herself in the moment, so that, when she was drawn back to memories of the war, she could reorient herself to her being safe at present. Eventually, L.C. was discharged from the Day Hospital but returned to the VA for group therapy sessions. Over time, her nightmares became less frequent, and she found that, even when something triggered a traumatic memory, it didn't necessarily send her back to the horror of that time and place. As a result, she began to feel more in control of her life. By the end of my internship at the VA hospital, L.C. was by no means "cured," but she was able to occasionally reclaim the sense of well-being that she had known before experiencing the trauma of war.

UNDER PRESSURE

Noradrenaline isn't the only neuromodulator that comes into play with emotion. In general, the brain systems in the central and peripheral nervous systems that regulate neuromodulators are highly interdependent, so no single chemical in the brain is associated with any particular emotion or experience. This is especially true of the brain's response to stress, an increasingly common feature of modern life.

In our daily lives, anxiety is a major driver of our stress responses. We feel anxious when we think something bad might happen, but we cannot predict or control whether it will happen. When we are under stress—say while working for a company that is laying off employees or caring for a loved one struggling with an illness that might get worse—several hormones are released, affecting everything from immune responses to glucose metabolism to neural plasticity. The hippocampus, prefrontal cortex, and amygdala all have receptors for stress hormones, and given the roles these areas play in our ability to recall information and experiences, it should come as no surprise that stress affects memory. That said, the effects of stress on memory are complicated; it sometimes enhances and sometimes squashes memory, depending on many complex factors that interact.

To study the ways in which stress affects memory, scientists have done some wacky experiments, from asking people to stick their hands in a jar of ice water to telling them that they will have to give an impromptu public speech or (my personal favorite) asking skydivers to study a set of gruesome photographs right before they jump out of the plane. All these manipulations elicit large increases in stress hormones. Cortisol is the most extensively studied hormone in these studies, and the spike that occurs when you are stressed-out can improve your ability to retain memories for what happened right before or right after the stressful event. Like noradrenaline, stress hormones seem to promote plasticity, initiating the cascade of changes that consolidate connections within the cell assemblies representing the memory for a stressful event.

It might seem odd that the human brain would have a mechanism that enhances memory for things that happened right *before* a stressful event, but it makes sense when you consider it from an evolutionary perspective. If you narrowly escaped being mauled by a wild animal, it would be important to remember not only that experience but also the circumstances that led up to it, so you can make sure not to repeat that mistake.

A lot of neuroscience research focuses on the memory-enhancing effects of acute stress in rats or mice, but stress has more complex effects on humans in the real world. Individuals differ hugely in their stress responses, for a variety of reasons, ranging from a history of anxiety or depression to past exposure to trauma, to the use of hormonal contraception, and even to insomnia. As is the case with pretty much any brain chemical, more isn't always better, but rather there are optimal levels for particular situations. Moreover, although neuromodulators can enhance retention of a memory for a stressful event, it doesn't mean we will remember it *accurately*. Stress tips the chemical balance in the brain, downregulating the executive functions mediated by the prefrontal cortex and enhancing the sensitivity of the amygdala. All of this means that, while you are under stress, you will perform poorly in situations requiring any kind of executive functions, and rather than learning with intention, you are more likely to have your attention (and hence memory) captured by whatever is salient. So, when you remember a stressful event, your memory is likely to emphasize your feelings and the factors you were stressed about, but you might have a hazy recollection of other aspects of the event. Additionally, if you are under stress, you'll have more trouble finding the right information when you need it. So, whether you're trying to recall your internet banking password or find your cell phone, you'll do yourself a favor by taking a deep breath and spending a moment doing whatever helps calm you.

Soldiers serving in combat, children living in an abusive home, or anyone who spends sustained time in a threatening situation or an environment dominated by a lack of control and predictability are especially vulnerable to the neurotoxic effects of stress and often show reductions in the volume of the hippocampus. This is particularly true for those suffering from PTSD and depression. Over time, the cumulative impact of stress may cause changes in memory and even contribute to the symptoms of PTSD. In the healthy brain, if the hippocampus does its job, then memories for traumatic events should be associated with a specific context. Hippocampal dysfunction in experimentally stressed lab animals, and in humans with PTSD, may cause traumatic memories to become overgeneralized. As a result, even situations or feelings that are only tangentially related to a traumatic event can bring back a flood of memories from the trauma. The combat veterans I worked with at the VA described how innocuous sounds, such as a car backfiring or the boom of a fireworks display, triggered intense flashbacks, even though these sounds were thousands of miles and decades away from the veterans' experiences in the Vietnam War.

In addition to PTSD, extreme stress can, in rare cases, trigger a disorder known as dissociative fugue. This disorder is frequently depicted as "amnesia" in Hollywood movies, but fugue is an extremely rare condition that is quite different from the kind of amnesia seen in patients like H.M. Individuals suffering from amnesia or who are in the early stages of dementia become forgetful, but they generally know their names and personal history and can usually recognize friends and loved ones. In contrast, people in a fugue state—such as the character Jason Bourne in the *Bourne Identity* books and movie series—have the confusing experience of finding themselves in an unfamiliar place with no sense of who they are. But unlike what you see in the movies, in dissociative fugue, sometimes the memories never fully come back.

A fugue episode is thought to explain the eleven-day disappearance of famed mystery author Agatha Christie. In 1926, Christie is said to have kissed her sleeping daughter before getting into her car and driving off into the night. The vehicle was later found abandoned, suggesting she might have been in a collision. At the time she went missing, the media speculated whether her disappearance was a publicity stunt to promote her latest novel or if her husband would become a prime suspect in a murder case. Then, in a plot twist worthy of one of her books, Christie was discovered at a hotel in Yorkshire two hundred miles north, where she had adopted the name and personality of her husband's secretary and had no memory of her own identity.

Fugue is often described as "psychogenic amnesia," as if to suggest that the condition is somehow unrelated to brain function, but the effects of stress on the brain are likely to play a significant role. In the few scientific studies that have been done on the disorder, the majority of patients appear to have experienced a significant life stressor, health problem, or neurological event prior to their fugue episode. In the curious case of Christie's disappearance, the devastating loss of her mother a few months earlier, followed by the revelation that her husband was leaving her for his secretary, as well as the possibility that she might have sustained a concussion shortly before abandoning her car, are compelling clues that point to dissociative fugue as a potential culprit.

Fortunately, most of us don't have the kinds of extreme experiences that can result in PTSD or fugue states. Still, we can see the consequences of stress in our daily lives—as when a fight with a loved one at home makes it difficult to concentrate at work—and there is good reason to believe it's important to manage and avoid the effects of chronic stress.

As social animals, our bodies' stress responses can easily be triggered by our interactions with others. The most reliable way to (ethically) stress out humans is to tell them that they are to be judged by other humans. Similarly, being in an unstable social hierarchy with a high level of unpredictability and uncontrollability can dramatically increase stress responses. Spending a long time in such an environment can have significant consequences for brain health, both indirectly (via effects on cardiovascular and metabolic functions) and directly (via long-term exposure to stress hormones). All this means that remaining stuck in an unpredictable, socially stressful situation, such as an emotionally abusive relationship or a toxic work environment, can be bad for your health and possibly detrimental to your memory.

SEX, DRUGS, AND DOPAMINE

Neuromodulators help us remember the highs as well as the lows. When I was a sophomore at UC Berkeley, a friend invited me to a birthday celebration in the dorms. I was sitting on the bed in his room when a lovely young woman with piercing blue eyes walked in and sat down beside me; soon we were deep in conversation. Her name was Nicole, and over a quarter century later, I still have a vivid memory of the emotional response I experienced upon meeting my future wife.

Dopamine, the neuromodulator that was likely spiking in my brain that night, is central to helping us form lasting memories for rewarding experiences. Whenever we experience something intensely rewarding, especially for the first time—a chocolate chip cookie fresh out of the oven, a sip of beer at an older cousin's wedding, kissing a pretty girl in a college dorm room—a burst of activity occurs in dopamine neurons. As a result, for many years, it was generally assumed within the scientific community that dopamine was responsible for the experience of enjoying a reward, and it's frequently referred to in the media as the "pleasure chemical." But the data tells a different story.

Dopamine is about *motivating* us to seek out rewards. Over many years, Kent Berridge, a professor of psychology and neuroscience at the University of Michigan, ran countless experiments showing that interfering with dopamine reduced the ability of animals to work for a reward, but it had no effect on their ability to enjoy rewards they received. In one study, however, Berridge flipped the script by setting up a rat in a cage with a metal rod that, when touched, delivered a small electric shock—not enough to traumatize or hurt the animal, but enough to induce a fight-or-flight response. He then used a laser to stimulate a part of the amygdala that activated dopamine circuits in the brain. Under normal circumstances, a rat will learn after one or two touches to steer clear of a stick that gives it a shock, but the rats whose dopamine circuits had been activated would approach and touch the rod repeatedly, even though they consistently got shocked. Berridge demonstrated that, to the brain, "wanting" something is not the same as "liking" it.

Like other neuromodulators, dopamine promotes neural plasticity, and its release tends to be concentrated in several brain areas that are important for helping us learn how to get rewards. Dopamine in the amygdala—the same brain structure that mediates fear learning—helps us learn about *cues* that signal an upcoming reward. In the hippocampus, dopamine helps us learn about the contexts where we are likely to get rewards. And finally, in a brain region called the *nucleus accumbens* (once thought to be the "pleasure center" of the brain), dopamine helps us learn what we need to *do* to get the reward. Because dopamine helps us form memories that associate cues, contexts, and actions that lead to rewards, it sets expectations that shape our experience when we get them.

You might think our brains would learn a lot every time we get a big reward, but in fact we are wired to learn only when the outcomes don't match up with our expectations. Dopamine activity ramps up when we are expecting a reward, and that expectation determines how the brain will respond to that reward. If we get a reward that is exactly what we expect, like a paycheck, there might be no change in dopamine levels. If we get a reward, but it is less than what we expected, like if our pay was docked, we might see a drop in dopamine levels, and if we receive more than expected, like a surprise bonus, we might see an increase in dopamine.

From studies on animals such as Berridge's rats, it appears that sudden drops in dopamine activity drain motivation, whereas an increase in dopamine activity can be energizing (though not necessarily pleasant). This means that a cup of coffee or an oven-baked cookie could get you excited, leave you with nothing, or be bitterly disappointing based largely on your expectations from past experiences. This variation in dopamine levels can keep us on a hedonic treadmill, sometimes working hard and joylessly just to escape that feeling of deprivation.

ACTING ON IMPULSE

One of the key lessons from research on the influence of survival circuits on memory is that these circuits do not just explain how we view the past, they can also influence our future decisions when we least expect it. It's easy to think that our choices are made willfully in the moment based on our knowledge of what we want to do, but in reality our choices are at least partly influenced by what our survival circuits want us to do, based on our past experiences. For instance, the research on dopamine and reward learning helps to explain why we often feel intensely motivated to seek out a reward even when we know it won't necessarily be pleasurable. When you find yourself cheating on a diet or giving in to a craving for a cigarette despite a New Year's resolution to quit, dopamine is driving the bus. We all have such moments when we make impulsive or even risky decisions, but some of us are more susceptible than others. Through an unusual set of circumstances, I studied what happens when we make risky decisions and why some people are more likely than others to do so.

When I joined the faculty at UC Davis in the fall of 2000, I had to hustle to get research grants so I could cover the costs of running a lab. Almost all biomedical research in the United States is funded by grants from the National Institutes of Health, but the competition for these grants is intense—in any given year, maybe one in ten applicants will get a grant. So, unsurprisingly, my first attempts to get NIH funding failed. Out of desperation, I scoured the web for alternative funding sources and found an obscure but lucrative opportunity from a nonprofit called the Institute for Research on Pathological Gambling and Related Disorders.

The timing was fortuitous. My first graduate student, Mike Cohen, was interested in understanding how we learn from rewards. We realized that gambling was a nice model for studying this process because even when the odds are stacked against the gambler, many people find that the thrill makes the risk worthwhile. After a couple of wins, some can be so compelled to persist that they lose large sums of money. By studying brain activity while people did simple gambling tasks, we realized we could better understand how the brain's reward-learning circuitry influences people's future decisions. We wrote this up in a grant proposal, code-named GAMBLOR, and got a grant on the first try. For two years, GAMBLOR funded my entire operation, including some studies of episodic memory along with the gambling studies we originally proposed.

Lab rats are typically rewarded with food or water, but humans can get a sense of reward from abstract events that have little survival value. If we record people's brain activity while they make the simplest of decisions ("Push the left or right button"), followed by the most minimal positive feedback ("Correct, you win!"), we can see increased activity in the nucleus accumbens, amygdala, and other brain regions where dopamine is released. In our gambling studies, we found that these neural responses weren't about the reward itself, but rather a sign of *learning* from the reward to guide future decisions.

As expected from what we knew of previous research in rats and monkeys, responses in the reward-learning circuit did not necessarily shoot up in response to the reward feedback, but rather in response to the extent to which the reward deviated from what was expected. When people unexpectedly won a bet, we saw a large neural response, but when they expected to win and did, we saw only a small neural response. If a person had a large neural response after seeing the outcome of the bet, they would be more likely to make the same bet next.

One of the unexpected discoveries in our studies was that people dramatically varied in their responses to rewards—we and others have seen this time and again. For instance, in one study people repeatedly chose between making risky or safe bets over an hour-long MRI scan. Those who played it safe were likely to win a small reward (80 percent chance of winning $1.25), while those who chose the high-risk decision were less likely to win a larger reward (40 percent chance of winning $2.50). Some people were systematically more likely to choose risky bets, and others tended to play it safe. The people who favored risky bets showed more activity in reward-learning circuits when they won than people who tended to play it safe. And unlike our risk-averse subjects, risk-takers still got a bump of activity in the reward circuit even when they lost a risky bet, and they were more likely to stick with a risky bet on the next trial. Mike's findings suggested that at least some people might have a reward-learning circuit that leads them to persist in making risky decisions even after bad outcomes.

Eventually, we spent the gambling grant, Mike graduated, and I got NIH funding to study episodic memory. It would be years before I would go back to studying rewards and learning, but that research changed the way I look at the choices we make in everyday life. Dopamine energizes us to seek out rewards but can also lead us to make rash decisions. Even being in contexts associated with the thrill of chasing rewards can be enough to make us act on impulse, whether it's cheating on a diet, getting blind drunk after reuniting with an old college roommate, having unprotected sex in a moment of passion, or driving recklessly in a fit of road rage. Later, in moments of calm, it can be hard to recall those contexts when our survival circuits took over. Our inability to remember how we felt when we were overcome with intense desire, fear, or anger can lead us to keep repeating the same mistakes. But if we can consciously remember the feelings that compelled us to make a bad decision, we can avoid putting ourselves in future situations that elicit that feeling.

This advice may be especially important for those who are prone to addictive behavior. As I noted above, some people are more likely to "learn" from such rewards and persist in chasing them even after bad outcomes. Just as some people are at greater risk for anxiety disorders such as PTSD that take over our brain's survival circuits for dealing with threats, others seem to be prone to the effects of drugs that hijack our survival circuitry for learning about rewards. Cocaine, methamphetamine, heroin and opioid painkillers, and alcohol all activate the dopamine system, sometimes driving the brain's reward-learning circuits much more intensely than any natural experience could. For some people, these effects are transient, but for others, the learning that takes place during drug use can lead to powerful addictions.

In my clinical work, I found that people with substance use disorders could stay clean for a significant time, but their recovery was often interrupted by environmental factors that brought them back into treatment. One of my patients, a charismatic storyteller named Rich who had served two tours in Vietnam, was struggling with an addiction to alcohol and crack cocaine. Despite the differences in our backgrounds and a twenty-year age gap, we formed an immediate bond over our shared love of soul and funk music from the sixties and seventies. Like L.C., Rich was haunted by memories of war, but the more we talked, the clearer it became that the real threat to his sobriety wasn't in his memories of the past but in the way his brain's reward system was being triggered by cues in the present: his friends were all still using, and he was living in the family home with his brother, who was a drug dealer. While Rich managed to stay clean for about nine months, he couldn't change his environment; seeing the same people and being in the same neighborhoods he associated with drug use would trigger intense cravings, and eventually he relapsed. Although many recovering addicts can maintain their sobriety long after treatment ends, for those such as Rich who find it difficult to break free from the cycle of addiction, it isn't just a question of will, it's a matter of memory triggered by context, and being in the wrong context can send them spiraling into relapse.

MANAGING THE IMPACT OF MEMORY

Pretty much every living thing on earth—even single-celled organisms—can learn to avoid threats and obtain rewards. But not all animals have episodic memory. Humans have both, and that combination helps us to survive in and navigate a complex world. The capability to revisit the past, combined with the brain mechanisms to learn about rewards and threats, helps ensure we retain the information that is most critical to our survival.

The discovery that we have separate episodic memory and survival circuits that interact with one another tells us a lot about both how we remember the past and how we use the past to make decisions in the moment. As creatures of memory, we gravitate to the highs and lows in our past. The bodily impact that comes with remembering past traumas can have a far-reaching impact in the present, overcoming us with emotions such as terror, shame, or anger that can lead us to make rash decisions we come to regret in the future. The sense of immediacy that comes when we are overtaken by our survival systems can make the emotional impact of memories seem impervious to reason or logic. Fortunately, we can learn to manage our responses from these systems.

My clinical work with individuals like L.C. was guided by the principles of cognitive behavioral therapy (CBT). The first component of CBT is challenging the interpretations that lead to negative thought spirals when we recollect emotionally charged experiences. As I'll describe later, the adaptive nature of memory can help us reframe the past, so we can gradually change our responses to emotional memories, while preserving our sense of what happened. The second component of CBT, which is equally important, is deliberately exposing patients to the physiological responses driven by survival circuits, so that those responses can gradually decline with experience. CBT is effective because its basic principles are backed by the science of memory.

Anyone can rely on these principles to manage the impact of remembering. We don't have to be prisoners of the past. Awareness is the first step. Once we are aware, we can work with our interpretations and the emotions that seem to come with our most intense memories, so that we do not allow our past traumas to define who we are in the present.

6

ALL AROUND ME ARE FAMILIAR FACES

HOW WE LEARN,
EVEN WHEN WE DON'T REMEMBER.

. . .

Right now, I'm having amnesia and déjà vu
at the same time. I think I've forgotten this before.

—Steven Wright

Endel Tulving's proposal that recollecting the past is an act of mental time travel was among the most significant and controversial ideas to emerge from the science of human memory. Although the idea remains controversial, the idea that episodic memory allows us to relive conscious experiences from our past now has a great deal of support. But we are influenced by the past in another way, by a force that typically works behind the scenes as we go about our business. At times, this force can manifest as a sense of *familiarity*, leading us to sense the existence of a memory even when it's not available to us. At other times, it can lurk beneath the surface of awareness, silently influencing our behavior in the present and shaping what we might do in the future. As we'll see, it forces us to bump up against deeper philosophical questions of human consciousness and free will.

Some years ago, before streaming platforms made meandering through aisles of DVD rentals on a Saturday afternoon a thing of the past, I was browsing in the comedy section of my local video store. I noticed that the young man behind the counter seemed familiar to me. Davis, California, where I live and work, is a small college town, so it would make sense that I had seen him before, but I had no idea where or when. He obviously felt the same way, as we repeatedly exchanged awkward glances while I thumbed through a shelf of DVDs. Suddenly, his expression changed, as if he had an aha moment, and he said, "Hey, you were the professor in my human memory class!" I then had my own aha moment and realized I had seen him many times before in my classroom, though I had failed to recollect anything about him.

Most of us have had a similar experience of someone's face giving us an overpowering sense of familiarity, leaving us scrambling for the memory responsible for that feeling. But why does this happen? How can we have a strong sense we've met someone if we can't remember anything about the person? It's not just faces that trigger this experience. Many of us have had a sense we've been in a particular place before, even if we know it's our first time there. Sometimes, we can even get a sense that an event unfolding in the present is one we've experienced before.

These shadowy memories emerge naturally from neural plasticity in the neocortex, that ability of our neurons to rewire themselves in response to something new. In this case, it is due to a process that exists to see and do things faster with less effort. But that sense of familiarity is only the surface of a powerful force shaping our behavior, for good and for ill.

DÉJÀ VU ALL OVER AGAIN

The term *déjà vu*, coined by a French philosopher in the late nineteenth century, translates into English as "already seen," but it's generally used to describe that peculiar sense of familiarity that can randomly occur while we are experiencing something new—such as when you arrive at a place you've never been and have the unshakable feeling that what you're experiencing in that moment, what you're feeling and thinking, has happened before. Déjà vu often manifests as this feeling of pastness, but it is also sometimes perceived as a premonition, as if a sixth sense were alerting you to what is going to happen next.

Déjà vu is an almost universal human experience, one philosophers, scientists, intellectuals, and artists alike have struggled to understand and explain for centuries. Plato and Pythagoras believed it was a residual memory from a past life. Sigmund Freud, the founder of psychoanalysis, argued that the sensation was a manifestation of our unconscious desires. His protégé Carl Jung theorized it was a product of the collective unconscious. In the canon of science fiction, déjà vu has been connected to supernatural phenomena ranging from time travel to multiple dimensions to alternative timelines. In the 1999 science fiction film *The Matrix*, it was portrayed as a "glitch" that happens when the code within our simulated reality has been altered. As compelling as these theories are, science tells a different story.

In the 1950s, Wilder Penfield, the pioneering neurosurgeon who co-advised Brenda Milner during her graduate studies at McGill, discovered that the feeling of déjà vu could be artificially generated. Penfield was trying to figure out a solution to a fundamental challenge in using brain surgery to treat seizures: removing the bad brain tissue without damaging any of the good. If you take out too little, the surgery does not alleviate the seizures, and the patient must return for another surgery. If you take out too much, a person could be left with major deficits in language, movement, vision, or memory (as was the case for H.M.).

To solve this problem, Penfield used an electrical brain-stimulation method developed during his training in Germany that is still used today. Surgeons would give patients a local anesthetic, perform a craniotomy (i.e., cut the skull open), then stimulate different parts of the brain with a small electrode, carefully observing behavior and asking the patients to report their experiences. If the stimulation caused a seizure, Penfield would know that this was a spot to remove. But other times, as Penfield stimulated different areas, his patients would report interesting feelings or sensations such as numbness in their fingers, flashes of light, or a strong odor. When Penfield stimulated areas in the temporal lobes, some patients reported a sense of déjà vu. One patient reported feeling "as though I had been through this before." Another described "a familiar feeling, very intense," and a third simply stated that "things seem familiar." Penfield's findings suggested that the brain can generate an intensely strong feeling of familiarity even if a specific memory does not come to mind.

About fifty years after Penfield reported that temporal lobe stimulation could elicit déjà vu in epilepsy patients, Rebecca Burwell, a neuroscientist at Brown University, found she could mimic this effect in rodents by stimulating a particular area of the temporal lobe called the *perirhinal cortex*. Rats, like human infants, tend to be more interested in exploring things that are new to them than in things they have seen before. Using an amazing technique called optogenetics, in which a tiny fiber-optic laser is used to activate neurons in a particular brain region, Burwell found she could manipulate whether an image seemed familiar to a rat. When she stimulated the perirhinal cortex at a high frequency, she found that her rats behaved as if an old and boring picture were new and interesting, and if she stimulated it at a low frequency, the rats behaved as if a picture they had never before seen were old and boring.

That electrical signals in the perirhinal cortex can artificially produce a sense of intense familiarity or novelty suggests this brain area might be responsible for that sense of familiarity we naturally experience when we visit a place or see a person we have seen before.

THE BEER BET

When scientific breakthroughs are depicted in the movies, it's usually as some brilliant scientist who has a "Eureka!" moment. We tend to mythologize the idea of "light bulb" discoveries—think Archimedes discovering how to measure volume after stepping into a bathtub or Sir Isaac Newton coming up with the law of gravity after seeing an apple fall from a tree in his mother's garden. That's not really how science works.

Most often, scientific progress comes about from the collective work of a diverse community. One researcher hears another's talk at a conference that was influenced by a paper written by someone else, which was inspired by a conversation with other colleagues while sharing a taxi to the airport. The research on familiarity in memory followed this trajectory, with scientific advances coming not from any one killer discovery but rather as a confluence of findings from many individuals using entirely different research approaches in different species.

For many years, the dominant school of thought was that memories fall along a continuum, ranging from "strong" to "weak." Adherents of this view would say that our sense of familiarity reflects a watered-down version of the memories from the hippocampus that, in a stronger form, gives us the sense of mental time travel. Someone from this camp would explain away my video store encounter by saying I just had a "weak" conscious memory for my former student. But this explanation didn't capture my experience. While I was certain the man was familiar, I could not recollect anything about him. Like Penfield's patients, I had the feeling of a "strong" memory, even though I couldn't grab the memory itself.

By the late nineties, neuroscientists such as the late Mort Mishkin, John Aggleton, and Malcolm Brown proposed that memory could be divided into different subcomponents, each of which could be strong or weak. This argument was based on the fact that animals with damage to the hippocampus, as well as the patients with developmental amnesia such as those studied by Faraneh Vargha-Khadem, seemed to do fine on "recognition memory" tests that required them to tell the difference between objects they had seen before and ones that were new. This work caught the attention of Andy Yonelinas, my friend and a longtime colleague at UC Davis.

Andy is from Canada, but with his long blond hair, incredibly mellow demeanor, and wardrobe of faded T-shirts, shorts, and sandals, you would think he grew up surfing on the coast of Malibu. Having worked with Endel Tulving as an undergraduate, Andy reasoned that episodic memory gives us access to a tangible memory from a specific place and time, leading us to feel confident that we are reliving a genuine moment from the past. In contrast, familiarity can be strong, such as the sense of certainty that you have seen something or someone before, or it can be weak, such as a hunch or educated guess; either way, it doesn't give us anything specific to hold on to.

In 1999, around the time Andy was presenting his ideas about disentangling episodic memory from familiarity, my postdoctoral mentor Mark D'Esposito was hired to start a brain-imaging center at UC Berkeley. Having spent about nine months at the University of Pennsylvania's neurology department, I moved with the lab back to the West Coast. Before settling in at Berkeley, I stopped to attend a neuroscience conference that was going on in San Francisco. There, at a "poster session," I first met Andy. Imagine a huge room in a convention center crammed with paper posters hanging on portable boards, not unlike a high school science fair, each fronted by a scientist excitedly explaining his or her latest discovery and fielding a gauntlet of questions from skeptical colleagues. Andy was presenting his research, and I might have been a bit too direct when I voiced my skepticism that familiarity was anything more than weak episodic memory.

I anticipated some blowback, but instead of getting defensive, Andy completely disarmed me by responding, "You should be skeptical!" After an afternoon of spirited debate, we decided to team up on an experiment to test the idea that we could identify a brain area responsible for the sense of familiarity. To sweeten the pot, we decided to make a "beer bet"—if his prediction turned out to be right, I'd buy him a beer and vice versa. (At this point, I've lost so many beer bets to Andy, it will take years to pay off my tab.)

We didn't yet have a working MRI scanner in the lab at Berkeley, but Mark finessed a space for us at a clinical MRI facility in the Martinez VA Medical Center, about halfway between Berkeley and Davis. The Martinez scanner was set up for routine clinical scans and didn't have state-of-the-art performance specs, so I had to MacGyver it, tweaking our procedures every way I could think of to turn this Ford Pinto into a Ferrari.

We were allowed to scan for free late at night after the clinic was closed. Over several nights, Andy and I hauled friends, students, and colleagues out to Martinez and scanned their brains while asking them questions about a series of words such as *lemon*, *wrench*, and *armadillo* ("Is it living or nonliving?" "Can it fit in a shoebox?"). We did this to give people a distinctive context for studying each word. For instance, if a subject imagined what it would be like to stuff an armadillo in a shoebox, that would create a distinctive episodic memory.

After pulling the subjects out of the scanner, we gave them a short bathroom break and then ambushed them with a surprise memory test for the words they had seen and for the context, asking for details about what they were thinking while they studied each word. Sure enough, activity in the hippocampus spiked when people saw a word and formed a memory that later helped them recall something about the context. But we didn't see this activity spike when people failed to remember contextual information, even if they were certain that they had studied those words before. We did not find any evidence that activity in the hippocampus was sufficient to give people a sense of familiarity, whether it was a hunch or a sure feeling they had seen something before. Instead, activity in the perirhinal cortex—the area of the temporal lobes associated with the déjà vu experiences in Penfield's patients and Rebecca Burwell's rats—was associated with memories that gave people a sense of familiarity. The more activity in the perirhinal cortex when our volunteers read a word in the MRI scanner, the more familiar that word would seem when they saw it again on the surprise test. And, unlike with the hippocampus, activity in the perirhinal cortex was not related to people's ability to recollect the context of words that they had studied. We had found evidence that memories are not just strong or weak; rather, the human brain has two different kinds of memory—episodic memory, which is supported by the hippocampus, and familiarity, which is supported by the perirhinal cortex.

I have never been happier to lose a beer bet.

Our beer-bet study was accepted for publication in late 2003, and our results attracted attention, but many scientists remained unconvinced. I had my own doubts—perhaps we had missed something some other lab might have found. To convince ourselves and the rest of the scientific community, we would have to see more evidence than just the results from our labs.

It was at this point that Howard Eichenbaum, director of the Center for Memory and Brain at Boston University, stepped into the picture. Howard had been researching learning and memory in rats and believed that a clue to understanding the neurobiology of human memory could be found by studying the functions of the rat's hippocampus. Unlike most other neuroscientists, he viewed his rats as sentient beings with an almost humanlike capacity for conscious thought. Endel Tulving had once described episodic memory as a uniquely human ability, so Howard, who loved to challenge the status quo, dedicated much of his career to showing that rats remember in many of the same ways humans do. So much so, his papers are famous for featuring irreverent, hand-drawn cartoons of pensive rats with thought bubbles depicting what was going through their minds.

A few years after the publication of our 2003 beer-bet study, Howard reached out to Andy and suggested that we team up and evaluate research from other labs all over the world—to put together results from studies of memory in humans, monkeys, and rats to see if familiarity is indeed different from episodic memory. A few months later, Howard was in Los Angeles for a conference, and we planned to meet up and begin our project. Andy and I took a quick flight from Sacramento to Los Angeles, hopped in a cab, and made our way to Howard's hotel. Howard was running late, so we passed the time chatting in a conference room, drinking weak hotel coffee and eating catered rubbery sandwiches Howard had ordered for us. When he arrived, we dropped the sandwiches and got to work. We soon got into a rhythm, the three of us bouncing ideas off one another in call-and-response fashion.

After a few hours, we had a solid plan for our project and returned home to dig through massive stacks of dense scientific papers reporting the effects of brain damage on memory in rats, monkeys, and humans, as well as a stack of fMRI studies of humans that had been published around the same time as our beer-bet study. None of these studies was, in and of itself, a smoking gun, but when we put it all together, almost all of them pointed to the same conclusion: familiarity isn't a weak form of episodic memory, but something else altogether, a form of memory that depends on the integrity of the perirhinal cortex.

Fifty years after Milner's first paper, we were able to bring together research across the entire field of neuroscience to answer the mystery of why H.M. had such a severe memory disorder. After his epilepsy surgery, he had lost both his hippocampus and the perirhinal cortex, so he could not rely on episodic memory or familiarity as a lifeline to his past experiences.

FLYING UNDER THE RADAR

We don't fully know how the perirhinal cortex contributes to our sense of familiarity, but we have good information about the learning mechanisms involved. Cell assemblies all over the brain are in con-stant flux, reorganizing and optimizing so that the neural elections that determine our perceptions, thoughts, and actions will come to swift and decisive conclusions. When those tweaks happen in sen-sory areas, they help us read, see, and experience the world more efficiently. Those little tweaks also happen in higher-level areas of the brain, such as the perirhinal cortex, that integrate information from our different senses to help build semantic memories.

All this neural plasticity seems to happen without our awareness, but the outcome can be sensed. The more familiar we become with something, the more our cell assemblies become fine-tuned to recog-nize that thing later on. So, if we pay attention to how much mental effort we put in to read a word or recognize a face, we can get a sense of how much experience we have with it.

For example, if I ask if you have ever eaten a rambutan, unless you grew up in Southeast Asia, you won't have to scrape through a lifetime of episodic memories to answer this question. There's a good chance that the work your brain put in just to read *rambutan* is enough to know it's unfamiliar. From that, you can infer you've probably had little or no experience with one of these hairy red fruits because you have not encountered the word often, if at all. You might also notice it takes a bit longer to access the meaning of the word *persimmon* than the word *apple*, and from that you can infer you have seen or eaten more apples than persimmons. In our fMRI studies, we have found that, when you initially think about a concept, such as a rambutan, activity spikes in the perirhinal cortex. It's as if this area of the brain is trying to match the word to a template you haven't yet created. In the aftermath of that encounter, the neural coalitions in this area get reorganized. The next time you think about rambu-tans, there's less activity because the election is resolved faster. The tweaking that happens after repeatedly seeing that word improves the brain's efficiency, reducing brain activity in the perirhinal cortex and making it easier to access the concept of rambutans.

At some moments, familiarity can give you an odd sense of aware-ness that you have a memory in there somewhere, despite having no direct proof that it exists—such as when the name of the actor in that series you just streamed is on the tip of your tongue, but you can't quite grab hold of it. When you think about that actor, enough action happens in the neocortex to give you some sense of familiarity, but not enough for the neural elections to be resolved. We are especially vulnerable to this "tip of the tongue" experience if we initially come up with the wrong name, as if the cell assemblies backing the wrong actor are suppressing votes from the neural coalition backing the actor you are looking for.

Familiarity can bubble to the surface in a way that gives us a sense of what we know, but it has a sneakier side that can indirectly influence our feelings and actions without our awareness. You've probably heard the phrase *familiarity breeds contempt*. The truth is, more often than not, most adults are actually drawn to the familiar—a phenomenon scientists call the "mere exposure effect." If you've seen or heard something recently, the fluency that comes from your prior experience can lead you to like it a little bit more the next time around. Sometimes, we can like it so much, we claim it as our own.

In *cryptomnesia*, sometimes referred to as "unconscious plagiarism," the brain mistakes a "forgotten" memory for an original thought or idea. Former Beatle George Harrison inadvertently learned this in 1970 when he penned the song that would become one of his biggest hits as a solo artist, "My Sweet Lord." Unfortunately for him, its success also led to a lawsuit when it became apparent that the music was uncannily similar to the song "He's So Fine" by the Chiffons, which was released almost a decade earlier. The publisher that owned the rights to "He's So Fine" sued Harrison for copyright infringement, and the case went to court in 1976. In his testimony, Harrison said that he was familiar with the song but insisted he had not consciously used it in his composition. After a protracted dissection of both songs, the court ultimately found him guilty of plagiarism.

The judge, Richard Owen, himself a classical musician and composer, summarized his opinion: "Did Harrison deliberately use the music of 'He's So Fine'? I do not believe he did so deliberately. Nevertheless, it is clear that 'My Sweet Lord' is the very same song as 'He's So Fine' with different words." Owen later noted that Harrison must have known that the melodies would work "because it already had worked in a song his conscious mind did not remember." Embittered by the arduous trial, Harrison expressed his displeasure with the verdict, writing in his memoir that he didn't "understand how the courts aren't filled with similar cases—as 99 percent of the popular music that can be heard is reminiscent of something or other."

The consequences of our plastic brains go much further than cryptomnesia. In our daily lives, we regularly have to rally our brains to deal with tough problems—questions with challenging answers, difficult decisions, and ambiguous situations. We might think that we deal with these problems based on a rational analysis of the information available to us, but our past experiences can subtly bias our choices. We often (mis)use familiarity as a heuristic, or mental shortcut, to guide decisions. Moreover, we can be blissfully unaware of these influences and instead reinforce our sense of free will by constructing stories that assign meaning to our choices and actions.

In a now classic paper published in the late 1970s titled "Telling More Than We Can Know," psychologists Richard Nisbett and Timothy Wilson set out to explore the extent to which we create narratives to make sense of our unconscious choices. In one experiment, student volunteers who had memorized the word pair *ocean-moon* were twice as likely to name Tide as the first laundry detergent to come to mind than the group of volunteers who did not see these words together. Things got more interesting when those students who had memorized *ocean-moon* were asked why Tide came to mind. None of them said, "Duh! I just studied the word *ocean*." Instead, they said such things as "Tide is the best-known detergent" or "My mother uses Tide" or "I like the Tide box."

If intelligent students who knew they were participating in an experiment could be so easily influenced without their awareness, consider the bombardment of advertising we are exposed to at sporting events, while watching television, browsing the internet, or even driving on the highway. Widely known brands such as Budweiser and General Motors aren't spending millions on advertising to make you aware of their existence or to rationally convince you why their products are better than the competition's. They are banking on the micro-influence just being exposed to their names could have on your choices.

Let's say you see an ad for Coca-Cola while watching the Super

Bowl. The company might have shelled out $6 million just to give you thirty seconds of exposure to its brand. But the infinitesimal tweaks in your brain that occurred over that thirty-second ad might increase your likelihood of grabbing a case of Coke by a tiny fraction of a percent. Multiply that by the 100 million people who watch the Super Bowl around the world, and that's a potentially huge increase in sales. Now consider that virtually the entire internet runs on advertising, from free music- and video-streaming sites to social media platforms and free email services.

A similar force is at play when we are moved to vote for a political candidate or donate to a particular cause that has been endorsed by a celebrity we think we know because they starred in our favorite TV series or movies. The familiarity that comes from this exposure endows them with an air of authority or expertise we believe we can trust, despite their having no other credential beyond their fame making them feel familiar to us.

FACE TO FACE

So far, we have considered how familiarity can be a decent indicator of memory, cuing us that we have seen something before. Unfortunately, it doesn't always work that way. As we found out in the last chapter, the neocortex resembles a neural network in the sense that it likes to find general patterns in the world, such as the features of birds. But what happens when we don't have a lot of experience with a particular category? For instance, I have little experience with flowers and can barely tell the difference between a carnation and a rose. We could simulate the same thing in a neural network. If it were trained to recognize only a few kinds of flowers, the network could end up treating all flowers as if they are the same. That's fine with flowers, but it has more significant implications when it comes to faces.

Consider the case of Robert Julian-Borchak Williams of Farmington Hills, Michigan. In January 2020, Mr. Williams returned home from work and was pulling into the driveway of his home when a police car suddenly pulled up behind and blocked him in. He was handcuffed in front of his wife and their two young daughters and taken to the Detroit Detention Center, where he was interrogated about a robbery that had taken place the previous October.

Even though Williams had a bulletproof alibi—he had posted a video on Instagram from his car at the time of the robbery—the possibility of an alibi wasn't factored in before the Detroit police sent out a patrol car to make the arrest. Why? Because Williams had been fingered by what was supposed to be an error-proof system. When the Detroit police ran a still image from a surveillance video of the thief through an automated facial recognition system—one that uses artificial intelligence to match up faces in photos with those in a database, such as driver's license photos—it flagged Williams as the culprit.

During questioning, a detective showed Williams the blurry image that had been used to identify him and asked if he was the person in the photo. "No, this is not me," he said, holding the photo up to his face. "I hope you guys don't think all black men look alike?"

Indeed, the image (which is publicly available) shows a black male with a large build but otherwise has little detail that could be used to recognize a face. Yet, even with such a poor-quality picture, it was clear that Williams was not the man in the photo. According to Williams, the investigating officers agreed, conceding, "I guess the computer got it wrong." Despite the obvious error, police kept Williams in custody for thirty hours after his arrest and released him on a $1,000 bond. Even after the charges were eventually dropped, the case was not dismissed until it was reported by *New York Times* journalist Kashmir Hill.

This is the first documented case in the United States of a wrongful arrest based on a faulty facial recognition match. Williams's experience makes a compelling argument against the use of AI face recognition systems as evidence in police inquiries, but it also shines a spotlight on the biases that exist in our brains and the society we live in. Automated systems learn to recognize faces through extensive exposure to massive sets of face photos, and these sets predominantly consist of white faces. I suspect that the designers of these algorithms were oblivious of the biases that they inadvertently embedded within their systems, but it is now a well-known fact that facial recognition technology disproportionately misidentifies or fails to identify the faces of Asians and blacks.

The problems with AI bias in law enforcement, however, pale in comparison to the problems caused by biases in human brains. Most police departments do not treat automated face recognition identification as solid evidence, but if a human eyewitness identifies the face of a perpetrator in a police lineup, it carries a lot more weight. Unfortunately, humans are notoriously biased at face recognition. A review of results from thirty-nine experiments (with data from almost five thousand participants) showed that, on average, people are 1.4 times better at recognizing the faces of people from their own race than those from other races, and that this bias is generally more pronounced for white observers than for nonwhite observers.

It's reasonable to think that human face recognition biases emerge from the same sources as those in computer programs. If you have been disproportionately exposed to people from your own race, when you meet another person from your race, your brain can leverage the expertise it has accumulated to rapidly put together the right coalition of neurons to identify that individual. Just as there is a critical period for speech recognition, our ability to recognize faces from different racial backgrounds might depend on exposure to faces from diverse racial backgrounds during that critical period before the age of twelve.

As we have seen from studies of chess experts, experience doesn't just change what you see, it changes *what you look for*. People tend to pay attention to the features that best distinguish faces of people from their own race, at the expense of features that help us recognize faces from other races. In his research on visual perception, psychologist Daniel Levin found that white observers were worse at recognizing the distinctive features of individual black faces relative to white faces, but they were better at recognizing the *race* of black individuals. Putting together all the evidence, it appears that we are more likely to form blurry memories of people from other races, in part because we don't get enough practice, and because we often pay attention to race over and above an individual's features. That means we can get a little sense of déjà vu for people of other races, even if we've never met them before—because our brains are doing a poor job of capturing what makes each face unique.

The societal implications of these biases go far beyond the problem of thinking that individuals from other races "all look alike." According to the Innocence Project, a nonprofit organization that has used DNA evidence to overturn hundreds of wrongful convictions, a significant proportion of convictions hinged on the testimony of eyewitnesses from another race. Fortunately, a growing movement now addresses these biases within the criminal justice system. In 2017, the New York State Court of Appeals issued a decision requiring judges to instruct jurors about the unreliability of eyewitness identifications in cases in which the defendant and the witness are or appear to be of different races.

All this means that, like every other aspect of memory, familiarity has both a good and a bad side. It can be a useful by-product of how the brain is constantly becoming more efficient in its perceptions, but its slippery nature means that the fluency that results from mere exposure to something can operate under the radar of awareness, influencing our choices, judgments, and behavior. When we go on autopilot, familiarity can constrain our options and leave us with a smaller world.

But autopilot is not inevitable. As we've learned, by using attention and intention, we can focus on remembering what is useful. What's more, we have the capability to tease apart the irrelevant impressions we bring from past experiences from information that is relevant at present. When we remember the power of familiarity to influence our behavior, we can get back a little bit of our free will.

7

TURN AND FACE THE STRANGE

HOW MEMORY ORIENTS US
TO WHAT IS NEW AND UNEXPECTED.

. . .

Curious is a good thing to be. It seems to pay some
unexpected dividends.

—Iggy Pop

When I was a kid, my favorite superhero was Spider-Man. I collected
the comic books and watched the animated series on TV after school,
but I would never have guessed that, twenty years later, Spider-Man
would go on to become a multibillion-dollar movie franchise. The
films have done an incredible job bringing to life most of Spider-
Man's extraordinary abilities, but in my mind his most interesting and
important superpower isn't one that requires any special effects; it's
his "spider-sense." Whenever danger was afoot, Spider-Man would
feel a tingling in the back of his skull, alerting him to scan his environment for a possible threat. This sixth sense functioned on a pre-
cognitive level, allowing him to spring into action even before he was
consciously aware of being in danger.

Although Spider-Man is a fictional character, there is something
to the idea that humans have a rudimentary spider-sense, in that our
brains can rapidly attune us to something that isn't quite right, even
before we're fully aware of it. I experienced one such moment when
I was an undergraduate at UC Berkeley.

I was living one block east of People's Park, which had been the
site of antiwar protests during the sixties, but had by then become
better known for drugs and muggings. My apartment was on the first
floor of a duplex—my roommate and I were fortunate enough to

have furniture left behind by the previous tenant, who had abruptly vacated after a nervous breakdown.

One afternoon, I came home and noticed a jacket lying on the ground in our driveway. I didn't think much of it; in that neighborhood, it wasn't unusual to find all sorts of things abandoned in our driveway. But when I unlocked and opened the front door, my spider-sense tingled. Something wasn't right. My eyes were drawn to the living room window, which was open, and then to the sofa, which had been moved to the side, and then to my CDs, which weren't as I had left them. My gut was telling me something was wrong, but I came up with an innocuous explanation—my roommate, Dave, who was active in the campus Young Democrats club, must have had a party. I went to pick up the phone (this was before cell phones were a thing) to call him, but the phone was gone. That's when it finally dawned on me that we had been robbed. I ran upstairs to use the neighbor's phone to call the police, and when I came back down to my apartment a few minutes later, the back door was wide open and the jacket in the driveway was gone—the intruder had still been in the house when I first walked in.

Looking back on this experience as a memory researcher, I wonder about that sense of unease I felt as soon as I opened the door. I may never know for sure, but based on our research, I believe the reason has to do with memory's most important feature, which is not about replaying the past but orienting us to the future. Our memories of the past—the "old"—enable us to allocate critical resources to what is new and what has changed. This capability likely played a central role in helping our ancestors survive in a volatile world. At a deeper level, it shows how memory allows us to transcend time and sense the connections between past, present, and future.

In the case of the robbery, I came through the door with a whole set of expectations and predictions: no jackets in the driveway, the window closed, the couch where I left it, the phone on the table—and definitely no thief in the house. When I stepped into my apartment, those predictions turned out to be wrong, and it triggered circuits in my brain that alert us to the unexpected.

Some would argue that the very purpose of memory is to predict the future. But it's not necessarily about making the right predictions. Sure, if we could always preload an accurate idea of what was about to happen, our brains would process that information efficiently. But those times we get it wrong—what neuroscientists call prediction errors—are important, too. Prediction errors initiate a cycle in the brain, in which memory (i.e., what we already know about the world) orients us to the unexpected, stimulating curiosity and motivating us to explore and resolve the gaps between our predictions and what we face in the present. The information we get while exploring, in turn, is prioritized in memory. As we will see, this cycle of predicting, orienting, investigating, and encoding lies at the heart of the universal human drive to learn and explore.

THE EYES HAVE IT

Our eyes move about four times a second. This mostly happens without our awareness, but these movements are anything but random. We know this from studies that use an infrared camera to track where people's eyes land when they are looking at a picture. Scientists used to think our eyes "fixate" on things that are salient, such as bright lights. John Henderson, one of my colleagues at UC Davis, who has spent his career studying what is likely to catch our eye, found that these factors actually play a minor role in the real world. Instead, as we go about our day, our eyes are directed by—you guessed it—memory.

First, we have general knowledge (semantic memory), which guides our exploration of the visual world, so we can find what is *supposed* to be in a particular place and rapidly identify when something is out of place. If you visit a friend's new house and she asks if you want to see the kitchen, before you even enter the room, your eyes are primed to head straight for the likely locations of things because you have some expectations, a schema based on your past experiences with kitchens. You might first direct your eyes toward the counter, where you would expect to see a coffee maker or a microwave. But if something violates your expectations, such as a blender on the floor or an empty space where a fridge would normally be, your eyes will immediately gravitate toward those spaces in the room.

Familiarity is another factor. We generally spend less time looking at things we've seen before, and more time looking at things that are novel. This is true of human adults and infants, as well as monkeys, dogs, cats, and rodents. It makes sense that our eyes do not linger on things that are familiar, given what we've discovered about how cell assemblies do not need to work as hard to process information about familiar objects, faces, or places.

Semantic memory and familiarity reflect what we've seen in the *past*, but memory does more than store knowledge or keep track of what we've seen; it points to what we can and should do in the *future*. In an incredibly prescient book published in 1978, Nobel laureate John O'Keefe and Lynn Nadel, Regents' Professor of Psychology at the University of Arizona, proposed that a major evolutionary function of the hippocampus is to tell us about places that are new or different, so we can explore and learn about these areas. Primates, and especially humans, explore the world with their eyes, so O'Keefe and Nadel's theory predicted that seeing something new or out of place should trigger a signal from the hippocampus stimulating us to explore our surroundings. Several subsequent studies have confirmed this prediction, showing that both monkeys and humans do more exploratory eye movements when seeing something novel than when seeing something old, and that this tendency depends on the integrity of the hippocampus.

My longtime friend Emrah Düzel, who directs one of Germany's largest Alzheimer's disease research centers, found that the hippocampal response to novelty might even be used to detect one's risk for the disease. In one study, Emrah's team had young adults look at photos of both novel and highly familiar places while they were in an MRI scanner. As expected, activity in the hippocampus increased when the subjects looked at pictures of new places—there was even an increase when people *anticipated* seeing a picture of a new place. These findings prompted Emrah to run a study to investigate hippocampal responses to novelty in a group of more than two hundred older adults. It turned out that older adults with high brain levels of tau and beta-amyloid— proteins that accumulate to toxic levels in Alzheimer's—showed a dampened hippocampal response to novel pictures, which in turn predicted poorer memory. Emrah's work suggests that the brain's response to what is new is tightly coupled to our ability to remember what we have previously encountered, and that a loss of this novelty response may be an early indicator of risk for Alzheimer's disease.

Finally, memories formed in the hippocampus play an important role in guiding visual exploration. For example, the first time you saw your friend's kitchen, your eyes might have been guided by semantic memory, which gives you expectations about the likely locations of objects generally found in a kitchen. But after that first visit, the hippocampus will help the rest of the brain make more precise predictions of where you should look and what you should look for. If everything is the same as it was on your last visit, the next time you walk into your friend's kitchen you'll make fewer, more targeted eye movements, and you won't be thrown off so much by that blender on the floor or the missing fridge. But if things are not where they are supposed to be, your eyes will quickly be drawn to those irregularities before you are even consciously aware something has changed.

In 2000, Jennifer Ryan and Neal Cohen at the University of Illinois published a clever study showing that the hippocampus is critical for attuning our attention to changes in our environment, much like the spider-sense I felt tugging at the hem of my awareness when I stepped through the door of my apartment in Berkeley. Ryan and Cohen used an eye tracker to study people's eye movements while they repeatedly looked at a series of photographs. As the photos became increasingly familiar, the volunteers didn't move their eyes around to explore as much. But, when Ryan and Cohen *altered* a picture—for instance, a photo of a child with a kitten in the background might be repeated, but with the kitten photoshopped out—the volunteers' eyes lingered on the spaces in the picture that had changed, and they kept returning to the same spot over several seconds. At times, they even fixated on empty spaces—areas that should have been completely uninteresting—because their brains caught that something was supposed to be there. Amazingly, even when subjects were unaware anything had changed, they spent more time looking at the altered areas, as if their spider-sense was tingling. In contrast, when people with amnesia were presented with an altered image, their eyes did not gravitate to the parts that had changed. Without a functioning hippocampus, their spider-sense was gone.

Why did people with healthy brains fixate on empty spaces, and why does this odd behavior depend on the hippocampus? When you go to a new place, or even look at a photograph of a new place, the neocortex can form individual memories of the little bits—the faces, the places, the objects. But as we have learned, we need the hippocampus to put together all the information about who, what, where, and when, so that later on, you can pull up coherent memories of the past. Ryan and Cohen's findings suggested that we might rely on these hippocampal memories to guide what to expect in the here and now. If something isn't in the right place, your spider-sense goes off, and your brain sends a message to your eyes to scan that area so you can figure out what happened.

Debbie Hannula, a postdoctoral researcher in my lab who had previously worked alongside Jennifer Ryan and Neal Cohen, set out to test this idea in an experiment that emulated how we associate the people we meet with the context of a particular place. She compiled a collection of photos of hair models, like the kind you might see in a shampoo ad or salon catalog. Volunteers in her studies encountered these attractive faces superimposed on images of particular places; for instance, they might see Trevor's face on a picture of a museum lobby or Mia's over the Grand Canyon. Debbie and I predicted that showing subjects in our experiments these familiar contexts again would lead the hippocampus to pull up a memory to generate predictions about who would be there, enabling them to quickly pick out Trevor's face after seeing the crowded lobby. To simulate this in the MRI scanner, we briefly flashed a picture of one of the places our volunteers had previously seen, and after ten seconds, we flashed three familiar faces on the screen and asked the volunteers to pick the correct face out of the crowd.

As each scene flashed on the screen, there was a large spike in activity in the hippocampus, consistent with the idea that seeing the place was enough to trigger retrieval of a memory from the hippocampus. This increase in hippocampal activity was enough to give them a heads-up, so that, when the faces were shown ten seconds later, their attention was immediately captured by Trevor and Mia, even in a crowd of competing faces.

In an interesting twist, the hippocampus was a little *better* at guiding people's attention than it was at pulling up a conscious episodic memory. Even when the hippocampus lit up in response to the museum, and a volunteer immediately fixated on Trevor, the volunteer sometimes still picked the wrong face. Picking the correct face depended on communication between the hippocampus (to make the right prediction) and the prefrontal cortex. Given the role of the prefrontal cortex in helping us to remember with intention, this finding made sense. The hippocampus, one of the most evolutionarily ancient structures of the brain, might be sufficient to point us in the direction of what is to come, as O'Keefe and Nadel proposed, but sometimes retrieving a memory from the hippocampus isn't enough unless we are able to *use* that information, which is where the prefrontal cortex comes in. Without that communication between the hippocampus and the prefrontal cortex, people were left feeling their spider-sense tingle but were unable to put their finger on what had triggered it.

WHAT IS IT?

I occasionally play in a cover band called Pavlov's Dogz, a collective of neuroscientists from all over the world. We usually meet at a neuroscience conference, and after spending two days rehearsing songs from the seventies and eighties—including songs by Blondie, David Bowie, the Pixies, Joy Division, Iggy Pop, the Ramones, and Gang of Four—we play a show to a packed house. Our band name was inspired by the Nobel Prize–winning Russian physiologist Ivan Pavlov, who is widely known for his foundational research showing that his dogs would salivate when they got a cue that predicted when they would receive food.

Cover bands and drooling dogs aside, I believe that Pavlov's most interesting contribution to the science of memory was to characterize our responses to novel and surprising experiences. He noticed that the animals in his experiments had a characteristic response to any changes or new things in the environment. He called this response the "What is it?" reflex, which I absolutely love because it so clearly depicts what's happening in our brains when we're confronted with something new or unexpected. Unfortunately, scientists could not leave well enough alone, so soon after Pavlov's work became known, they came up with a more opaque and sterile term.

The *orienting response*, as the "What is it?" reflex ultimately came to be known, is an orchestrated set of changes in the brain, and indeed all over the body, in response to something new or surprising. Our pupils enlarge, increasing sensitivity to light. Blood is pumped to the brain and constricted in the rest of the body, and the brain gets a brief shot of neuromodulators, such as dopamine, noradrenaline, and acetylcholine. There's also a coordinated change in neural activity throughout a network of brain areas, including the hippocampus and the prefrontal cortex.

The orienting response is triggered when we deal with unexpected or surprising events. Its ubiquity in a variety of species is an indicator of the evolutionary value of identifying and responding to the unexpected. The simplest way to measure the orienting response in humans (or any mammal) is to record electrical activity in the brain while someone listens to a series of *beep* sounds, punctuated by an occasional *boop*, at which point the subject is told to push a button. If you do this task, you'll quickly learn to expect an unremarkable stream of beeps and boops, and all a neuroscientist needs to do to activate your brain's orienting response is to insert an incongruous sound, such as a duck quacking or a dog barking. At the beginning of the experiment, a huge brain wave occurs every time someone hears one of those incongruous sounds. But by about twenty minutes into the experiment, those curveballs are so mundane they no longer elicit much of a brain response. We eventually come to expect the unexpected.

Two of the recurring characters in this book—the hippocampus and the prefrontal cortex—are key players in generating the "What is it?" reflex. We know this through recordings of activity in the hippocampus and the prefrontal cortex when people encounter surprising stimuli, and from studies of humans with damage to these areas.

Back when I was helping to set up the brain imaging center at UC Berkeley, I heard an impromptu talk by a visiting German epileptologist named Thomas Grunwald. Following the method popularized by Wilder Penfield over half a century earlier, Grunwald was using electrodes to record information from different brain areas to locate where seizures were taking place, so that he could remove the dysfunctional brain tissue without damaging intact areas of the brain. When individuals with a healthy hippocampus listened to a sequence of sounds, Grunwald observed a large electrical spike in the hippocampus in response to surprising noises. But that response was virtually absent in patients with hippocampal dysfunction. Grunwald found that the orienting response is probably one of the most reliable indicators of a functioning hippocampus and thus was a useful tool to determine when the hippocampus needed to be removed to stop a patient's seizures.

I was intrigued by Grunwald's findings, which seemed to suggest something important about why some events seem to be especially memorable. We already knew that the ability to form new memories is amplified when we're motivated by a threat or reward—as Mike Cohen and I had seen in our gambling study—and we wondered if a similar effect happens in the brain when we orient to something unusual. So, Thomas Grunwald, Nikolai Axmacher, an up-and-coming neurologist at the University of Bonn, Germany, and I devised a study to figure out how the orienting response might be related to learning. Mike was inspired to get involved with this study. So he secured a prestigious fellowship from the German government, packed his bags, and moved to Bonn to start recording directly from the human brain.

Nikolai's clinical work involved evaluating patients who had electrodes implanted in their brains in order to find the source of their seizures. While the doctors were waiting to record seizure activity from their brains, the patients had a lot of time to kill, so they were willing to participate in our study. In our experiment, volunteers were asked to memorize a set of pictures of faces and houses. In some picture sets, a face would be shown amid a stream of houses or a house would be shown amid a stream of faces. We expected the oddballs to elicit some kind of orienting response, which could affect people's ability to remember those images. To analyze if these responses were related to memory, we initially recruited individuals with epilepsy who, like Penfield's patients, had electrodes implanted in different parts of the brain to identify where their seizures were coming from, and later we looked at orienting responses in those who had relatively normal activity in the hippocampus.

In science, things often become more interesting when an unexpected connection is thrown into the mix. Not long after he arrived at Bonn, Mike ran into a scientist who was using a radical approach to treat individuals struggling with severe depression. The theory was that these people were suffering because of a dysfunction in their brain's reward system, so the clinical team sought to "jump-start" the system by implanting electrodes to stimulate the nucleus accumbens, a brain area that uses dopamine to learn about rewards. After implanting the electrodes, surgeons recorded brain activity from the electrodes (to verify that they were in the right brain area) before turning on the stimulators. Mike was excited about the possibility of recording electrical signals directly from the brain's reward-learning circuitry and convinced the team to invite these patients to participate in our experiment while the surgeons were recording brain activity during the electrode implantation procedure. Mike had a hunch that dopamine might play a key role in helping us learn about things that are surprising—and if so, we should be able to discover an orienting response in these patients.

So thanks to Mike and Nikolai, we were able to recruit two groups of patients, one with electrodes in the hippocampus and another with electrodes in the nucleus accumbens. Our recordings from the first group showed that less than two hundred milliseconds after one of the surprising faces or houses flashed on the screen—barely enough time to make an eye movement—a blip of activity occurred in the hippocampus. These results fit with the idea that the hippocampus is like a "What is it?" detector, alerting the brain to something unexpected. We also found a second spike of activity in the hippocampus a little over half a second after an oddball image flashed on the screen, and this increase predicted whether the subject would be able to remember the surprising item later on. These results suggested that the hippocampus was also preferentially forming memories for the unexpected oddball pictures.

When we looked at the data from volunteers who'd had stimulating electrodes implanted in the nucleus accumbens, we discovered that Mike's hunch was right. Activity in the accumbens spiked about half a second after they saw an oddball image. We knew that activity in the nucleus accumbens is associated with increases in the release of dopamine in the brain, which mobilizes us to get rewards. But external rewards couldn't explain our results. In our study, there was no payoff for seeing and memorizing an unexpected house shown amid a bunch of faces or a face in a stream of houses—the oddball pictures were just as relevant as pictures from the expected category. Our findings suggested that surprising or unexpected events can be sufficient to trigger activity in this system even when we do not get an external reward.

Putting together the results from the epilepsy patients with electrodes in the hippocampus and the depression patients with electrodes in the nucleus accumbens, our results supported a theory proposed by the late Brandeis neuroscientist John Lisman, which described how surprising events trigger a sequence of neural responses that increases learning. According to his theory, the hippocampus initially responds to surprising events, which then mobilizes the nucleus accumbens into action, signaling areas deep in the brain to release dopamine, which opens the floodgates for the hippocampus to form new memories for the surprising information.

As we've already seen, not all memories are equally important. Just as our brains preferentially capture memories for the highs and lows, they also prioritize learning about stuff that is surprising or new. That makes sense because when you encounter something you fully expected, such as a blender on a kitchen counter, there isn't much reason to memorize it. But if you saw a chain saw sitting on your friend's counter, that would probably be something worth remembering (and perhaps a source of concern).

The results from our study also pointed to a new direction my lab would take in the next few years. As we have seen, dopamine *energizes* us to seek out rewards to satisfy that motivational drive. If dopamine circuits are recruited when we make prediction errors, such as when we encounter an unexpected event, I began to wonder if prediction errors could also energize us to seek out information. We would soon find out the neural circuitry that we had been investigating was fundamentally related to the drive to learn.

CURIOUS CREATURES

Pavlov described the orienting response as a *reflex*, meaning that it is an innate biological response. As he eloquently described in a 1927 lecture:

The biological significance of this reflex is obvious. If the animal were not provided with such a reflex, its life would hang at every moment by a thread. In man, this reflex has been greatly developed with far-reaching results, being represented in its highest form by inquisitiveness—the parent of that scientific method through which we may hope one day to come to a true orientation in knowledge of the world around us.

Pavlov believed that the orienting response, which is seen in all sorts of animals, is foundational for something much bigger, perhaps even fueling humanity's loftiest achievements. Our greatest works of art, literature, and philosophy, all the discoveries we've made as we continue to explore new frontiers of science and the known universe, may well come down to the link between how we respond to the unexpected and the intrinsic drive to find answers. Pavlov called it inquisitiveness. Most of us would simply call it *curiosity*.

But what exactly is curiosity? What makes us seek out answers to the unknown, and what is the point of having a brain that is motivated by curiosity? The psychologist George Loewenstein, a pioneer in the field of behavioral economics, argued that curiosity is triggered when we discover a discrepancy between what we know and what we'd like to know, a nebulous space he called an "information gap." Loewenstein proposed that curiosity doesn't actually feel good but instead is an unpleasant state akin to thirst or hunger that compels you into action to get some relief—like an itch you need to scratch. Just as Kent Berridge found a difference between the motivation to get a reward and the pleasure we get from it, Loewenstein proposed that curiosity is about the motivation to *seek* information, rather than the satisfaction of getting our questions answered.

It's not hard to understand why our brains might be wired to get basic rewards (e.g., food, water, comfort) that ensure our survival, but why would we have a similar drive to seek information? Neuroscientists have argued that this drive is evolutionarily adaptive because it helps us maintain a balance between *exploration* and *exploitation*. Consider what would happen if our cave-dwelling ancestors had no drive to explore beyond their immediate surroundings. When they were hungry or thirsty, they might have randomly roamed around their environs until they found places to get water and food. They could then simply exploit what they had already learned and continue visiting those foraging sites again and again, but as these resources become scarcer, they might encounter competition and have to fight over that space. Things could get ugly.

Now imagine cave dwellers with curiosity. Motivated by a drive to explore, they might venture away from their known territory to explore what lies beyond the tree line or over the next hill. Perhaps they find a new and better foraging site, or they might come face-to-face with a venomous snake. Whatever happens, they are certain to acquire information. That information has value because it widens their knowledge of the world around them.

The nature of curiosity, once a fringe topic in cognitive neuroscience, has blossomed into an entire field of research, bringing together scientists interested in motivation, decision-making, and memory in both humans and nonhuman primates. One of the most intriguing findings to come out of this inquiry is that there might be a trade-off between curiosity and external rewards. Rewards can sometimes reduce our intrinsic motivation to do a task or follow our interests. Conversely, both monkeys and humans are willing to forgo external rewards in exchange for information needed to satisfy their curiosity. For example, neuroscientist Ben Hayden conducted an experiment with rhesus macaque monkeys given a video gambling task. He found that the monkeys in his experiment did not enjoy being in suspense, so they were willing to settle for lower winnings (a smaller amount of juice or water) in exchange for finding out sooner whether they had won the bet.

The trade-off between chasing tangible rewards and chasing knowledge is as true in the real world as it is in experiments. I can speak from personal experience: the emotional, intellectual, and financial investment in pursuing a PhD may be the ultimate example of forgoing rewards in exchange for the opportunity to satisfy curiosity.

Although curiosity is about the motivation to learn, only recently have we begun to investigate whether being curious actually *improves* learning. I came to understand more about this possible connection between curiosity and memory through a freshly minted PhD from

University College London, Matthias Gruber, who joined my lab in 2007 as a postdoctoral researcher. Having done his PhD in England, Matthias was well acquainted with the British obsession with pub quizzes. The tradition, in which teams of bargoers compete in a live trivia competition, began in the 1980s as a way to attract customers on off nights. Since then, it has become as much a part of British culture as fish-and-chip shops (aka "chippies") and Sunday roasts; according to some estimates, something like a third of the United Kingdom's roughly sixty thousand pubs run at least one weekly quiz.

The goal of competing in a pub quiz is to show off your knowledge and beat your opponents, but a huge part of the appeal is also the inherent allure of trivia questions. Sometimes you get a question ("What's the world's largest land mammal?") and the answer comes to you immediately (the African bush elephant), so you feel satisfied with yourself. Other times, a question comes up ("What year was Marmite invented?") and you have no clue about the answer (1902) and no interest in finding out, so you get impatient for the next question. But every once in a while you get one you should know the answer to ("What was the first rock single to hit number one on the *Billboard*'s leading pop singles chart?") but you don't. Those are the questions that make you curious. (And for those who are, the answer is "Rock Around the Clock" by Bill Haley & His Comets.)

Matthias's arrival in my lab coincided with the publication of an innovative fMRI study by Min Jeong Kang and colleagues at Caltech. Kang proposed that when you get stumped by a trivia question you *should* be able to answer, what you're experiencing is an information gap—a kind of prediction error—that triggers the brain's motivational circuitry to energize you to find the answer. The study found that activity in a dopamine-receptive area of the brain increased in response to trivia questions that piqued the curiosity of their subjects. Moreover, the participants were much better at remembering answers to questions that piqued their curiosity. Kang and colleagues proposed that, by tapping into the brain's reward circuitry, curiosity can enhance memory.

Matthias was intrigued by this idea and approached me about designing experiments to find out more about the link between curiosity and memory. Initially, the topic seemed frivolous to me, but

Matthias was persistent, and eventually I set aside my skepticism and encouraged him to follow his own curiosity.

If dopamine energizes us to seek out rewards, we wondered, what if it also energizes us to seek out information? And if dopamine motivates us to learn, then maybe the boost we get from curiosity isn't from getting the *answer* to the question we're interested in, but rather from the *question* itself. To test this idea, Matthias designed an ingenious experiment. Pulling together a large pool of trivia questions, he prescreened them with the participants and asked whether they knew the answers. If they didn't, he asked them to rate how curious they were to find out. Next, our volunteers hopped into the MRI scanner, where we showed them the questions that had stumped them, then delayed showing them the answers for ten seconds. While they lay there waiting in suspense for each answer, they were shown a picture of a face. When they got out of the scanner, we tested them on the trivia questions and noted the answers they got wrong or didn't know.

To no one's surprise, our volunteers remembered more of the answers to questions that triggered their curiosity than those they were uninterested in. The more surprising result was that they were also better at remembering the faces they had seen when they were in suspense, waiting for the answers to questions that stimulated their curiosity. The faces had nothing to do with the trivia questions and were, by any account, completely uninteresting. Yet, once our volunteers' curiosity was stimulated by a trivia question, being in a curious state helped them learn these faces that they *weren't* curious about.

Our fMRI results helped to explain what was going on. When people saw a trivia question they were curious about, activity increased in the dopamine circuits in the brain (including the nucleus accumbens). More activity in these circuits in response to the question predicted more curiosity to find out the answer. Just as Loewenstein argued that curiosity is about *wanting* information, dopamine circuits seemed to be triggered by the *questions* that made the participants curious, rather than by learning the answer.

The second key result was that, although most people showed better learning when they were curious, some did not. Those who did get a learning benefit from curiosity showed increased communication between the hippocampus and the dopamine circuit. Our findings suggest that, when a question stimulated curiosity enough to get a shot of dopamine to the hippocampus, people could also take in information that they weren't particularly motivated to learn.

Chasing information to satisfy our curiosity can sometimes have bigger effects on memory than learning in order to get an external reward. In the years since we published our first curiosity study, several labs have run similar experiments showing that curiosity improves memory for both the mundane and the interesting, whether you're eight years old or eighty-eight. In contrast, external rewards seem only to enhance memory for information that we are not curious about.

Putting together all our findings from the effects of novelty, surprise, and curiosity on memory, the big takeaway is that the brain's reward circuitry isn't about rewards per se; it's about mobilizing us to learn and pursue anything we perceive to have value. That could be food or water, but once these basic needs are met, what is the most significant thing we need to chase? *Information.*

CHOOSING CURIOSITY

So far, we have explored the cyclical relationship between memory and exploration. We use memory to make predictions. Prediction errors energize us to gain new information, which will be prioritized in memory. There's only one problem—it doesn't always work out that way. In Matthias's trivia studies, we saw a wide range of variability in the degree to which curiosity enhanced learning. Some people showed a big advantage in learning when they were curious, while others seemed to get no learning boost. We are now trying to figure out how and why the effects of curiosity on learning and memory differ across individuals. Some of that variability might lie in temperament.

Our findings suggest that the people who show the biggest learning benefits from curiosity also score high on a personality trait called *openness to experience.* This fits with other work showing that openness to experience is a better predictor of learning even than simply being interested in academic pursuits. People who rate high in openness also tend to be receptive to unconventional ideas, have an appreciation for diversity of beliefs and cultural practices, are will-ing to explore new places and topics of interest, and enjoy learning in the absence of a particular goal or achievement. More research needs to be done on the temperamental contributions to curiosity, but it is reasonable to think that all of the above behaviors reflect a fundamental human drive to seek information.

That said, I don't believe temperament alone dictates curiosity. When something is unexpected, it can motivate us to explore. If that uncertainty makes us uneasy about the world, the unfamiliar can also be scary. When I was a child, travel made me anxious. Being at hotels, staying at relatives' houses, or even sleeping over at a friend's house was stressful. I wasn't scared of anything in particular. My fear was of the *unknown*—falling asleep in a new bed, waking up in an unfamiliar house.

As an adult, I've learned to enjoy the exhilarating energy that comes with travel. Despite my inability to sleep on transatlantic flights or the frustration of struggling in high-level scientific discus-sions with some of the greatest minds in my field while my brain is addled by jet lag, I crave the stimulation that comes from being in a new place, seeing new things, and meeting new people. It feels like an extended version of the orienting response, with my pupils dilated, heart rate elevated, and eyes darting all over as I take in unfamiliar sights. But my excitement can still easily shift to anxiety under the wrong circumstances. I still have trouble sleeping in hotel beds, in part because of that nervous energy of being in a new place.

I suspect, to some extent, we all have a complex relationship with what we don't know or understand. According to one survey, one in six Americans have never left the confines of their home state. When we find ourselves in unfamiliar territory, we are pushed out of our comfort zone. Some are troubled by any deviation from the routine and find the presence of outsiders to be disturbing. It speaks to a fundamental choice we make, consciously or unconsciously, about whether to respond to the unknown with curiosity or anxiety.

Anxiety and curiosity seem like opposites, but the hippocampus seems to play a role in both responses. Most of the evidence for hip-pocampal involvement in anxiety comes from studies of rats with damage to the front part of the hippocampus, which is nestled just behind the amygdala. Oddly enough, these animals act much less fearful than healthy rats—they're much more adventurous in trying new foods, socializing with strangers, or exploring new places. Not having a fully functioning hippocampus seemed to make the animals more likely to venture into the unknown.

How is it that damage to the hippocampus, an area of the brain that appears to be essential for generating orienting responses that prompt us to explore and learn, can make an animal *more* willing to explore and learn? The clue to solving this mystery is that the hippocampus allows us to generate expectations based on past experience. When we're in a new place, meet a new person, or encounter something surprising, there's a conflict between what we thought would happen and what we are confronted with. Like Peter Parker's spider-sense, sensing a prediction error is enough to alert us that something is amiss, but not enough to tell us what's happening or what we should do about it. That information gap may be what triggers the circuits in the prefrontal cortex that enable us to use our knowledge and goals to determine the next steps. According to Paul Silvia, a psychologist at the University of North Carolina at Greensboro, information gaps can trigger curiosity and exploration if one feels one has the capability to close the gap. If not, that information gap can seem more like a chasm because we simply don't know how to go about closing it. Or it can be scary—an indicator of an uncertain and potentially threatening situation.

This brings us back to one of the recurring themes in memory: the complex mechanisms of the human brain that have evolved over the millennia can be good or bad, depending on how we use them. Memory helps us notice disruptions in familiar patterns. It could be a jolt triggered by a loud noise as you walk through an unfamiliar neighborhood, the feeling that schoolchildren experience when trying to solve a tough math problem, or a nagging sense that your significant other is acting unusual and may be hiding something. At these moments, we have a choice. We can mobilize our defenses and withdraw, we can avoid the situation by telling ourselves that nothing is amiss, or we can let our motivational circuits energize us to explore and plunge ahead with curiosity and *turn and face the strange*.

PART 3

THE IMPLICATIONS

. . .

8

PRESS PLAY AND RECORD

HOW REMEMBERING CHANGES OUR MEMORIES.

· · ·

Those aren't your memories.
They're somebody else's.

—*Blade Runner*

A popular trope in science fiction is for a protagonist to travel back in time and either accidentally or deliberately change the past, typically with dire consequences. Just as physical time travel could result in feedback loops that alter the present or future, a similar principle applies to the mental time travel we do when recollecting a past event. If you revisit the past and introduce something new to the memory, the past can be changed. At least, your memory of it can be changed, which for your brain is more or less the same as if you had traveled back in time and changed the past itself.

In 1906, Hugo Münsterberg, chair of the psychology lab at Harvard University and an outspoken researcher on the fallibility of memory in criminal prosecutions, received a letter from Dr. J. Sanderson Christison, regarding the case of a young man named Richard Ivens, who had confessed to a brutal murder. The victim, Elizabeth Hollister, had been strangled with a copper wire, her body dumped several hours later in a trash heap four blocks from her home on the outskirts of Chicago. Her jewelry and purse were missing, suggesting robbery as a possible motive.

Richard Ivens had been caring for his father's horse in a nearby barn when he discovered Hollister's body. Although he had no prior criminal record or history of violent behavior and no evidence tied him to the murder, he became the prime suspect in the investigation. Ivens initially denied any involvement, but after hours of relentless questioning, during which police pressured him to recall committing the crime and at least one officer brandished a gun, he ultimately broke down and confessed—twice. The first confession was terse and had many gaps. The second, a written account of the murder taken immediately afterward, was far more detailed, though his recollection still had a number of inconsistencies. Ivens was then questioned by the assistant state's attorney, and his answers largely corroborated his written confession.

Newspapers all over the country pounced on the story, running front-page articles proclaiming Ivens's guilt, but the case was riddled with holes. During the trial, a number of witnesses, including his parents, friends, and other family members, provided credible alibis suggesting that Ivens could not have been the murderer. Even the state's attorney noted that, according to the time line provided by Ivens himself, he could not have committed the crime. Moreover, when interrogated, Ivens's initial story of his actions during the murder contradicted key details of the crime (only to be later corrected with information fed to Ivens by his interrogators).

Six days before his scheduled execution, Ivens asserted his innocence and denied having any memory of even confessing to the crime:

> From the time I was arrested I do not believe that I was myself for a moment. . . . I suppose I must have made those statements, since they all say I did. But I have no knowledge of having made them, and I am innocent of that crime. . . . I know that the very first thing that the Inspector said to me when I was brought to him was "You did this." I did not do it, and I knew that I did not; but I do not know what I said or did during that time in the station.

Hugo Münsterberg, who had been conducting pioneering research on the inaccuracies of human memory, was drawn into the case, and unwittingly into the public eye, by the letter from Dr. Christison. After inspecting the evidence in the case, Münsterberg concluded that Ivens was indeed innocent and replied to Dr. Christison that Ivens's confession was "exactly like the involuntary elaboration of a suggestion put into his mind . . . a typical case of that large border-land region in which a neurotic mind develops an illusory memory." Münsterberg's letter, which was leaked to the press, set off a fire-storm of newspaper headlines ("Harvard's Contempt of Court," "Science Gone Crazy") and was the opening salvo in a fierce debate that would continue well into the next century about the malleability of memory.

Despite Ivens's having recanted, and the extensive arguments for his innocence made by Münsterberg and a number of other prominent psychologists, the jury took only thirty minutes to find him guilty of first-degree murder. Ivens was hanged a month later—less than five months after he had found Elizabeth Hollister's body.

It might be difficult to fathom why an innocent man would confess to a murder he didn't commit, but sadly Ivens's case is not unique. The annals of justice are riddled with cases of individuals who experienced convincing recollections of their involvement in gruesome crimes they did not commit. Which begs the question: How could someone *remember* committing a crime if they did not do it?

Consider it an unpleasant side effect of mental time travel.

Our memories are not etched in stone; they are constantly changing as they are updated to reflect what we have just learned and experienced. Although it might seem counterintuitive, the catalyst for memory updating is the very act of remembering. When we remember, we do not passively replay the past. Accessing a memory is more like hitting "play" and "record" at the same time. Each time we revisit the past in our minds, we bring with us information from the present that can subtly, and even profoundly as in Ivens's case, alter the content of our memories. Consequently, every time we recall an experience, what we remember is suffused with the residue of the *last time* we remembered it. And on it goes; each step is one link in a neural chain subject to edits and updates, so that, over time, our memories can drift further and further from that initial event.

A COPY OF A COPY OF A COPY . . .

When professor of psychology Elizabeth Loftus of UC Irvine, whose study of eyewitness testimony I discussed in chapter 4, was fourteen years old, her mother drowned in a swimming pool. Decades later, at a family gathering for an uncle's ninetieth birthday, a relative brought up the subject of Elizabeth's mother's death and insisted that she was the one who had found the body. This came as a surprise to Elizabeth—as she remembered it, her aunt had found her mother floating in the pool. Yet, this family member was positive it was her. In the days that followed, Elizabeth repeatedly traveled back to that day in her mind, and hazy images began hovering like ghosts at the edges of her memory. *Her mother floating face down in the water . . . Firemen arriving on the scene and pressing an oxygen mask to her face.* Memories lining up with what the family member had told her began to bubble up, and Elizabeth started to think maybe she *had* been the one to find her mother after all.

A week later, the relative called to apologize. Other family members had confirmed it was indeed Elizabeth's aunt who had found the body. Loftus was shocked. Not because her brain had conjured up vivid details of something she hadn't witnessed, but because *she* of all people should have known it could happen. To this day, she is arguably one of the most well-known—and controversial—memory researchers working in the field. So, how was it that a memory of such a traumatic moment in her life could so easily become corrupted?

The answer, which Loftus discovered in her own groundbreaking studies, is that we are especially vulnerable to *misinformation* at the moment of remembering. In one classic experiment, Loftus had people watch a series of slides showing a car stopping at an intersection with a stop sign. After the slide show, the participants completed a questionnaire, and some of them were asked, "Did another car pass the red Datsun while it was stopped at the yield sign?" It was a trick question. What they saw in the slide show was a stop sign, not a yield sign. Yet that tiny bit of misinformation worked its way into people's memory of the event. One week later, the majority of participants in her study remembered seeing a yield sign.

This type of memory distortion played a key role in the corruption of Loftus's memory of the circumstances of her mother's death. Elizabeth had received misinformation from her relative, leading her to repeatedly attempt to recall the traumatic scene. In imagining what could have happened, her memory became increasingly unmoored from the original event, to the point that she nearly became convinced of the imagined event.

Even when we aren't exposed to misinformation, repeatedly accessing particular memories can lead to their being updated in more subtle ways. Frederic Bartlett, who described recollection as an "imaginative reconstruction," conducted a number of studies that showed how our memories can be changed by remembering. In his "Experiments on Remembering," Bartlett had students at Cambridge study a series of invented pictograms or read a short story (as he did in the "War of the Ghosts" study). Then, on successive occasions, he repeatedly asked them to reproduce what they had learned. Bartlett observed that his subjects could initially recall a great deal about what they had learned, but over time, their reconstructions drifted from the source material. With each subsequent recollection, idiosyncratic details were lost, and inaccuracies began to creep into the narrative.

When I think about everyday memory updating, I'm reminded of my college days, playing gigs with a punk rock band at pubs and bars around Berkeley. To promote our shows, we would make flyers and post them all over town. We wanted them to stand out, so we would type out the information about our show on an old electric typewriter, then photocopy the text at the maximum level of enlargement. Then we'd enlarge and copy the copy, then enlarge and copy that copy, and on and on until the typed letters were huge. With each copy, the text got larger, but it also became more and more distorted—little imperfections in the typeface, tiny flecks of dust on the glass surface of the copier, and ink smudges on the printer became larger and more prominent. The result was pretty cool, almost like an artifact from a vintage printing press. The fundamental information we had typed out the first time remained the same, but its look and feel had changed.

When we revisit a memory over and over, subtle alterations can creep in with each repetition. It's a bit like making a copy of a copy of a copy; the neural connections that hold together a memory are tweaked, and these changes can enlarge some aspects of the experience, while causing us to lose some of the details that keep the memory in focus. Like the fuzzy letters on my college band flyers, events from the distant past can seem more remote and blurry every time we call upon them, and noise becomes more prominent, corrupting the memory a bit more each time it's recalled.

To understand why remembering can introduce distortions or lead our memories to become less detailed, a number of scientists in my lab have worked with a computer model of the hippocampus that was developed by my longtime friend, and one of the world's most well-known computational neuroscientists, Randy O'Reilly. We recruited Randy in 2019 to join the UC Davis Center for Neuroscience, and over the years he's become my go-to partner on many stand-up paddling, wake surfing, and occasional skateboarding expeditions (though he has become a little more careful since fracturing his wrist in an accident while riding a motorized one-wheeled skateboard). After poring through paper after paper on the specific types of neurons in the hippocampus, the way in which they are connected, and the physiology of how these connections change with learning, Randy put together a painstakingly detailed computer model of the hippocampus that allows us to simulate what happens when we learn new information and when we subsequently try to recall it. We have used Randy's model as a sandbox to play around in and explore how memories are formed, retrieved, and updated, and it has been an extraordinary resource, leading us to new and sometimes counterintuitive insights about the nature of human memory.

Our simulations show how easy it is for memories to become corrupted as we revisit them over and over. Just as the cell assemblies in our brain are constantly reorganizing when we struggle to learn something new, our modeling suggests the same thing takes place in the hippocampus when we recall something from the past. Suppose you see something that reminds you of your first date. The first time you recall this memory, the hippocampus might pull up information about the context, sending you back to that summer evening as you stammered as you said goodbye at the end of the night. Each time you remember that event, your hippocampus faces a problem: At the moment of remembering, you are not in the same place or the same mental state you were in back when that first date took place. In our computer simulations, we found that the hippocampus is effective at catching the difference between past and present and updates the memory accordingly. Every time you recall that first date, cell assemblies in the hippocampus reorganize, incorporating features of what's happening at the time you are trying to reconstruct the experience. If, at the moment of remembering, you imagined a scenario in which you were less awkward, a little of this information might get incorporated into the memory. Every time you recall the event, the memory updates a little bit more, accumulating more and more updates, so that you have effectively traveled back in time and turned your awkward teenage self into a silver-tongued charmer.

Our modeling gave us insights into how memories can gradually become altered with each retelling. But what is happening when a memory is entirely false, when an individual has the experience of a memory, yet that memory has no connection to an actual experience?

THREE TRUTHS AND A LIE

In 1980, Michelle Smith and her therapist (and later husband), Dr. Lawrence Pazder, triggered what would come to be known over the next decade as the "satanic panic" with their book, *Michelle Remembers*. Smith recounted how, over hundreds of hours of hypnotherapy sessions with Pazder, she had recovered suppressed childhood memories of horrific trauma she had endured as a prisoner of a satanic cult decades earlier. Although Smith had "recovered memories" of several gruesome murders and human sacrifices, including the butchering of stillborn babies and being fed the ashes of the victims, none of the claims were corroborated by her sisters, and extensive investigations revealed no evidence for any of the crimes described in her book.

Nonetheless, *Michelle Remembers* inspired a generation of therapists to use Pazder's techniques, leading hundreds if not thousands of individuals to develop vivid memories of extreme "satanic ritual abuse," of which they had no memory prior to entering therapy. Over the next decade, law enforcement agencies across the United

States opened criminal investigations into suspected satanic cults, with Pazder serving as a consultant for the prosecution.

As these cases gained media attention in the 1980s and into the 1990s, the spotlight fell on memory researchers such as Elizabeth Loftus, whose studies were called on to make a scientific case for refuting the "recovered memory therapy" techniques popularized by Pazder and others. Until then, research on misinformation and memory updating had gone largely unnoticed by the rest of the world. But this epidemic of criminal investigations based on recovered memories ignited a public debate—dubbed the "memory wars" by author and literary critic Frederick Crews—between experimental psychologists studying the fallibility of memory and mental health practitioners who specialized in the treatment of trauma survivors.

Having already shown that memories can become corrupted by misinformation planted at the moment of remembering, Loftus wanted to see how far her experiments could go. It seemed possible that, in at least some cases, recovered memory therapy might actually cause people to develop memories for events that had never taken place. So Loftus wondered if this phenomenon could be produced in the lab. Would it be possible to implant a *new* memory? To get someone to construct a false memory entirely from scratch?

She had a moment of inspiration at a party where she told a friend about her memory implantation idea. Her friend ran with it, asking his daughter, Jenny, "Remember the time you were lost?" Loftus questioned Jenny and probed her about whether she'd been frightened. In their conversation, as the young woman began to recall fragments of an event she had never experienced, Loftus saw firsthand that it might be possible to implant a memory, and she set about putting together an ingenious recipe to bring her idea to fruition.

Loftus drew from factors she had identified in her previous research, including repeated attempts to recall a past event, imagination, and time. But her experience with Jenny led her to the last and possibly most crucial variable: misinformation from a trusted source.

Working with her research assistant Jacqueline Pickrell, Loftus devised an experiment in which participants read stories about events from their childhood that were compiled by a trusted close relative, such as a sibling or parent. Three of these events actually happened, but one story, about being lost in a shopping mall, was fabricated. Still, knowing that the stories came from a trusted relative, the participants had no reason to suspect one of them was false. In two subsequent sessions, from one to two weeks apart, the participants were asked to write down anything they could recall about each of the four events. Initially, they recalled little to nothing about being lost in the mall, but by the first interview session, "memories" started to emerge, and even more details were added by session two. At the end of the experiment, Loftus and Pickrell revealed that one of the events was made up. When asked which story they thought was false, most of the participants singled out being lost in the mall, but a sizable minority—five out of the twenty-four—picked a real event as the one that was manufactured by the experimenters.

Loftus and Pickrell's experiment might seem artificial, but it is remarkably faithful to the real-life circumstances that help explain why so many people in the era of the satanic panic developed vivid yet entirely false memories (as even Loftus did with the false memory of discovering her mother's body). Repeated attempts to remember merged with misinformation from a trusted source are a potent combination that can lead people to construct a memory born not of experience but of *trying to remember*.

The original "lost in the mall" experiment has been criticized for a number of reasons, but since that time, a large number of well-controlled studies have substantiated Loftus's original conclusions. An analysis of results from a number of studies that used her memory implantation recipe showed that, on average, about one in three people can become convinced that they experienced an event that never occurred, ranging from childhood memories about an accident at a wedding, to being a victim of a vicious animal attack, to nearly drowning before being rescued by a lifeguard, or even witnessing demonic possession as a child.

As Loftus herself acknowledges, implanted memories such as being lost in a mall are not entirely false; rather they are most likely imaginative constructions that incorporate information from schemas and details from real experiences. After repeatedly attempting to recall events that must have taken place, participants start to pull in what Loftus and Pickrell eloquently describe as "grains of experienced events" along with a healthy dose of imagination. Eventually, memory updating from repeated retrieval attempts can lead the grains of truth and imagination to coalesce into convincing memories that are difficult to discriminate from memories of events that actually took place.

Not everyone succumbs to memory implantation, but certain elements can make it more likely to work. Age is one factor. Young children and the elderly are, on average, more vulnerable to memory implantation, as are those with a tendency to dissociate (i.e., tune out the outside world). Guided imagery techniques such as hypnosis, in which one is encouraged to imagine a hypothetical scenario, or the use of alcohol or drugs also significantly increase vulnerability to memory implantation.

Ironically, the ingredients for Loftus's recipe for memory implantation are essentially the same as those used in some forms of recovered memory therapy, the goal of which is to help a patient "recover" memories of a traumatic event. Recovered memory therapy involves repeated suggestions from a trusted individual (i.e., the therapist), visualizations of how the traumatic event could have occurred, and the occasional use of hypnosis or medications. Moreover, recovered memory therapy is often directed at individuals who are prone to dissociation. Proponents of this type of therapy claim that the fictional events that are implanted in memory studies do not resemble the kinds of traumatic memories reported by their patients. But unlike lab studies, which involve a few sessions with a researcher who is effectively a stranger, a patient undergoing recovered memory therapy forms an intense relationship with a therapist over an extended time with the shared goal of pulling out memories for traumatic events. Recovered memory therapy is like the Loftus memory implantation recipe on steroids.

Fortunately, the effects of memory implantation can be reversed if an experimenter tells participants that they might have been misinformed or encourages them to be skeptical about the accuracy of the memories they produced. These findings from studies in the lab are consistent with real-life examples. For instance, one individual who recovered memories of satanic ritual abuse in therapy eventually saw a new therapist and came to believe she had inadvertently repackaged different past experiences into recollections of events that never happened. In the months that followed, she traced her memories of satanic ritual abuse to a variety of other sources: a movie she'd seen as a teenager, a short story she'd written, and the popular 1973 book *Sybil*, about a patient with dissociative identity disorder.

That people can differentiate between real and implanted memories suggests that, even when we update our memories, we have the capability to monitor what we recall to discern fact from fiction.

FALSE CONFESSIONS AND MISINFORMED WITNESSES

Hugo Münsterberg, whose letter in defense of Richard Ivens was met with a public rejection of his scientific research, was a passionate advocate for using psychological research and the science of memory to improve the criminal justice system. His studies raised important questions about the validity of cases based solely on the memories of eyewitnesses and confessions made under duress. Research on misinformation and memory implantation has since validated Münsterberg's concerns, and these findings have far-reaching implications.

One study, for instance, investigated whether Loftus's memory implantation recipe, including repeated prompts to recall an event, misinformation, and suggestive questioning by authority figures, could lead unwary participants to develop false memories of having committed a crime. As in the "lost in the mall" memory implantation study, participants' parents provided information about real events, and the experimenters added a fictitious event: a crime the subjects had committed that resulted in police intervention. Over multiple days, the participants were repeatedly asked to recall the event (and instructed to try to imagine how it could have happened, if they did not remember). At the end of the experiment, over 25 percent generated rich personal memories of a crime, and another 40 percent believed that they had committed a crime.

Interrogation procedures in the real world can elicit the same kinds of effects. The "Reid technique"—an interrogation procedure introduced in a 2001 manual widely used to train detectives—outlines a series of tactics that would be fully expected to contaminate a suspect's memory. As in the Richard Ivens case, suspects are first told that they are guilty of a crime, which alone could lead them to doubt the accuracy of their initial attempts to remember what happened. Suspects are then repeatedly presented with a selective set of facts about the case and asked to imagine how the crime had unfolded. Sometimes the suspects are deliberately shown false incriminating information, which can further lead them to doubt their memory for the events in question. Interrogators are instructed to present themselves as sympathetic to the suspects to gain their trust.

Hours of repeated attempts to recall an event under extreme duress, combined with misinformation from a potentially trusted authority figure, provision of key details of the crime, and coaching to imagine how the crime unfolded can lead a vulnerable individual—such as Richard Ivens—to recall a crime that the person never committed.

Law enforcement procedures can also corrupt the memories of eyewitnesses. Eyewitnesses are vulnerable to memory updating when leading information is provided—sometimes unintentionally— while they are attempting to recall a key event. Consider the case of Jennifer Thompson, who was sexually assaulted in 1984. Working with police to identify the perpetrator, Thompson was asked to help generate a composite sketch. She looked through several pages of facial features—noses, eyes, eyebrows, lips, ears, and so forth. Next came a photo lineup. Perusing the photos, Thompson identified a young black man named Ron Cotton as a potential suspect, though she was unsure. After she picked out Cotton, the detective told her, "We thought this might be the one."

Thompson was then asked to identify the suspect in a lineup. Standing in the same room with the man she suspected of assaulting her made Thompson sick with anxiety. She was unsure which of two men in the lineup was the perpetrator, but eventually she settled on Cotton. As Thompson walked out of the room, still shaken from the experience, police told her, "That's the same guy. . . . That's the one you picked out in the photo." Thompson was confident that she had identified the perpetrator, and in 1985, Cotton was now convicted of the crime. After spending ten years in jail, Cotton was exonerated when the real perpetrator was conclusively identified by

DNA testing. Thompson, having undergone one trauma, had been retraumatized by procedures that created the ideal conditions for a catastrophic case of memory updating.

Seeing the finished composite sketch after she attempted to remember the perpetrator's face was the first step in a chain of events that affected her memory. During the photo lineup, she attempted to recall the face of her attacker, and she received confirming feedback—which turned out to be convincing misinformation—from the police. Adding even more complications, Cotton was the only person who was in the photo lineup as well as the physical lineup, thus Thompson might misidentify him based on her updated memory from the photo lineup. When Thompson reported she was still unsure, police took the final step of providing more misinformation by confirming that she had indeed made the correct choice.

The case of Jennifer Thompson and Ron Cotton is one of many in which repeated suggestive questioning of eyewitnesses and interrogation of suspects or eyewitnesses under extreme duress has led to memory distortions with disastrous consequences. The law enforcement community is becoming increasingly aware of the scope of the problem, and many agencies have adopted reforms based on scientific research on memory updating to prevent such injustices.

MALLEABLE, BUT NOT MUSH

Some years ago, I was approached by a U.S. attorney who was prosecuting a landlord accused of sexually assaulting a number of women who lived in his buildings. She explained that it had taken years for the victims to become aware of one another, and now that they had come forward, she was concerned that the defense was going to use Loftus's research to argue that, after all this time, the victims' memories were no longer reliable.

Unfortunately, trauma survivors have become collateral damage in the memory wars. Media articles on the malleability of memory often extrapolate wildly from studies of memory updating, suggesting we can *never* trust anyone's recollection of events that took place long ago. This attack has consistently been used against those who have come forward in the #MeToo movement. That's a dangerous claim. Not only does it sow self-doubt and confusion in the minds of survivors of traumatic events, it's inconsistent with the science.

If you look closely at the literature on misinformation effects, you can see that memory is malleable, but it isn't mush. Scientists who study memory accuracy and the legal ramifications (including Elizabeth Loftus) are well aware of the limitations of false memory effects in the lab. As I mentioned above, at least one analysis of laboratory studies of memory implantation suggests that the majority of people do *not* recollect rich false memories of an event that never occurred. And even though misinformation can corrupt memories, under some conditions, misinformation can help people become *more* accurate. For instance, if people remember the original event well enough to catch the misinformation, they can reinforce their initial recollection. As we will see in the next chapter, our ability to strengthen memories that we have just recalled is a fundamental feature of human learning.

It's also important to understand that memory-updating experiments rely on a specific recipe of misinformation implanted by trusted sources, and that is not representative of the experiences of many trauma survivors. Most survivors of childhood sexual abuse do not have to recover traumatic memories. On the contrary, the memories are all too often consistently available to them. Survivors can be remarkably accurate at recalling the pertinent aspects of their traumatic experiences years, and even decades, after the abuse occurred, and contrary to the idea of repressed memories, severe trauma can sometimes be associated with more accurate memories. Like many of the veterans I worked with at the VA hospital who suffered from service-connected PTSD, survivors tend to struggle not with a sudden recollection of forgotten traumas but with the nightmare of constantly reliving those experiences.

Even though scientists have not found evidence to support the concept of repressed memories, there are plenty of reasons why people might seem to have a "recovered memory" (i.e., the experience of remembering an event and believing it was not available to them before). This might happen simply because they never really forgot it in the first place. In many cases, survivors do not disclose their traumatic experiences until later in life. An outside observer might erroneously conclude that it's a "recovered memory" and therefore unreliable, when in fact the memory was there, but the survivor was previously not ready to share the experience with others. In other cases, a survivor might not have thought about the traumatic event for a long time and eventually lost track that it had been available in the past. Later on, being in a particular context or situation, as when we revisit a location where a trauma occurred, might remind us of an incident that we had otherwise forgotten.

Perhaps the most interesting aspect of memory updating is the degree to which forgetting or accurate remembering is within our control. There's no scientific basis for the idea that we automatically repress traumatic events, but some evidence suggests we can suppress memories in a way that is not too different from suppressing an urge to grab a candy bar while waiting in line at the supermarket. Researchers have simulated this phenomenon in the lab by having people memorize words, then after they've cued the memory, they are told to suppress the memory and not think about it. Remarkably, this simple instruction works. Actively trying to keep something out of mind can make the memory less accessible later on, even if the subjects are offered money to recall the suppressed words.

In studies of college students studying memory for words or pictures, the effect of suppressing memories isn't huge, but it could play out in a more significant way for some people in the real world. Remembering traumatic experiences is painful, and understandably many survivors actively try to keep the memories out of mind. Over time, suppressing these memories over and over might lead them to become a little blurrier and harder to recall.

All this research has clarified the circumstances under which we can trust our memories and those in which we should be concerned about the possibility of memory updating. It has also helped us move beyond the memory wars toward a more nuanced view of memory. At this point, most clinical psychologists accept that repeated suggestions can lead to memory updating and distortion. Conversely, most experimental psychologists (myself included) would agree that people can accurately recall many aspects of traumatic events and that memory updating rarely leads people to form completely new memories for extreme or traumatic events. And we all can use this emerging understanding to avoid the negative consequences of memory updating and instead harness its power to improve our lives.

THE UPSIDE TO UPDATING

Our brains aren't wonky. They are wonderfully adapted to make use of the past, given the dynamic and unpredictable world in which we have evolved. Although it's clearly not adaptive to remember events that never took place, having mutable memories of the past has important advantages. The world around us is constantly changing, and it's critical to update our memories to reflect these changes. When the restaurant you used to love is taken over by new management and suddenly gives you food poisoning, you need to make an adjustment in your dining preferences. If you catch someone you thought was trustworthy in a lie, you're going to infuse some skepticism into the next piece of information you receive from that person. Without memory updating, we would lack the flexibility to adjust our behavior based on new information.

Many neuroscientists believe that memory updating is a function that's guided by our genes. As we touched on in chapter 5, when we have a new experience, connections between neurons are modified, enabling us to access the memory later. Neuromodulators help these changes to solidify by turning on genes inside the neuron, instructing the cell to manufacture proteins that help to increase the strength and efficiency of the connections that are forged during learning. Researchers used to think that, a few hours after learning, this process is more or less done and the memory solidified—the technical term for this process is *consolidation*. That dogma was upended when neuroscientists Karim Nader and Joseph LeDoux published research showing that the sequence of steps in consolidation seems to happen whenever we retrieve a memory.

Suppose animals learn about a threat—say, by a small shock upon hearing a beeping sound—and the consolidation kicks in, stabilizing the memory for that association. Later on, if you remind them of the tone and don't shock them, the memory becomes destabilized.

By stopping the cells from making those proteins, you can effectively erase that bad memory, so that the animal seems completely unfazed by the tone when it is played later on. The conclusion from this study and others seems to be that, when we recall an event, we have to reconsolidate it, and if reconsolidation is blocked, we can erase the memory.

The most reliable approaches to disrupting reconsolidation can only be used in nonhumans. For instance, you can study mice with genetic mutations or use some fairly potent (and somewhat toxic) drugs to disrupt the molecular changes that kick in after a memory is retrieved. Obviously, it's not ethical or even feasible to use these approaches in humans. Scientists have attempted to demonstrate reconsolidation in humans in other ways, ranging from distraction techniques to administering fairly harmless drugs that block fear responses when someone is reminded of a frightening memory. These studies have had varying degrees of success, and different labs have had trouble replicating one another's results. Despite these challenges, there have been a number of clinical trials of reconsolidation-based treatments for PTSD, and researchers are exploring whether the effectiveness of psychedelic-assisted psychotherapy might be due to destabilization of traumatic memories by drugs such as MDMA.

In my mind, the jury is still out as to whether reconsolidation can reliably be harnessed in the clinic, but a deeper message is to be gained from the research. We already know that memories can be strengthened, weakened, or modified from the moment they are pulled up. This kind of memory updating is at the heart of psychotherapy, which is fundamentally about changing connections that we made in the past in the face of new information. The goal isn't to erase someone's memories of what happened but to adaptively update memories and change one's relationship with the past by approaching it from a different perspective.

This doesn't apply solely to traumatic memories. Many of us remember everyday experiences that are laced with unpleasant feelings, memories of those we've wronged or who have wronged us. Having a brain that has evolved to allow us to revise these memories means we can reframe how we feel about them by incorporating a new perspective. When you recall your boss snapping at you, perhaps you can consider reframing the memory after you consider that he was having an intensely stressful day. Or a memory of a bad date can be seen as an opportunity to learn about what qualities you are looking for in a partner. If we can harness remembering to reframe the way we look at the past, we can update our painful memories so that we experience them in a more tolerable way, possibly leading us to growth.

9

SOME PAIN, MORE GAIN

WHY WE CAN LEARN MORE
WHEN WE MAKE MISTAKES.

. . .

I learn from my mistakes. It's a very painful way
to learn, but without pain, the old saying is,
there's no gain.

—Johnny Cash

One of the unique advantages of being both a memory researcher and a college professor is that the work unfolding in my lab informs the work I do in the classroom and ultimately makes me a better teacher.

I taught my first undergraduate class, Human Learning and Memory, in 2002. My daughter, Mira, was an infant, and I had just started at UC Davis, as an assistant professor. Although I was a bit green, the class went well, and I continued to refine it over the next eighteen years. By the time Mira grew up and moved away to college herself, I had settled into a comfortable routine in teaching my class. That is, until a global pandemic forced me to reconfigure my approach to teaching.

In January 2021, after ten months of lockdowns and social distancing protocols that forced educators to rapidly pivot to virtual instruction, I was dreading the start of a new quarter. Like millions of teachers around the world, I was facing the challenge of how to teach and inspire a grid of faces on a computer screen. Having watched my daughter, Mira, struggling to stay engaged in courses that she would have loved to take in person, I couldn't help but worry about how my students were taking to online instruction. I realized

I had to radically rethink how I taught this course if I was to rise to the challenges of remote learning.

For close to two decades, I had been basing my students' grades primarily on the results of two midterms and a final exam. It's easy to administer tests in person, but with online instruction it's nearly impossible to prevent cheating, so I had to let go of the traditional idea of using tests as a yardstick to measure achievement. Once I surrendered to this reality, it opened me to the possibility of using testing for an altogether different purpose. Rather than continuing to think of tests as a tool for measuring what was learned, I decided instead to use testing as a tool to *drive* learning. To do that, I had to look no further than the work we had been doing in my lab.

We were doing brain imaging studies based on decades-old research showing that testing people on recently viewed material dramatically increases the ability to retain that information over time. Doubling down on the model we were using for these experiments in the lab, I switched to a format in which my students had a three-day window to take an open-book quiz online every week. Soon after the completed quizzes were submitted to me, students could see the correct answers and learn from their mistakes. If they got it right, seeing the correct answers reinforced what they had learned. The point of these weekly quizzes wasn't to torture my students but to prompt them to think critically about the material we had covered in class and to give me an opportunity to provide feedback and support when necessary. The topics covered on the quizzes were revisited on the midterms, such that students could use each quiz as an opportunity to sharpen their knowledge in anticipation of the midterm.

Midway through the semester, I had my students fill out an interim course evaluation to see how well we were doing in the transition to online instruction, and the response I got exceeded my wildest expectations. What amazed me the most was that when my students were asked "Did the weekly quizzes help you learn?" 85 percent responded "strongly agree." When asked about the best part of the class, one even said, "The quizzes, because they help me prepare for the midterm and being tested on things that I just learned is great for retention." If you are not a teacher, let me assure you that students almost never say anything positive about any kind of test. Yet, in this case, they had described my new weekly testing format as one of the best parts of the course.

In retrospect, I don't know why I didn't adopt this method years earlier, because it exploits a simple principle of human brain function. We are wired to learn from our mistakes and challenges—a phenomenon called *error-driven learning*. I believe that this simple principle can explain a wide range of phenomena—the conditions under which we learn best, the conditions that lead us to forget, and possibly even the changes that happen in memory while we sleep.

THE STRUGGLE IS REAL

The year I turned fifty, I made two resolutions: the first was to write a book—the one you are reading—and the second was to learn how to surf. The latter has been more humbling than any challenge I encountered on the journey to get my PhD. Part of the problem is that I'm out of shape, and surfing puts incredible demands on core muscles I didn't even know I have. The real difficulty, however, is that there is so much to learn: paddling technique; "reading" the waves so I can get in the right position and start paddling at just the right time; figuring out when I have truly caught the wave; and—if I am lucky enough to get to this point—popping up to my feet before I nose-dive into the water. Any mistakes along the way lead to a wipeout, which is both physically taxing and demoralizing.

Whether you're taking up a new sport such as surfing or learning to speak a new language or play a musical instrument, the pain from making mistakes is one of the biggest obstacles to learning, and it's especially frustrating when you are early in the learning curve. But even an expert who seems to perform such tasks naturally was once a novice who started off by making mistakes. Novices progress to becoming experts by continuing to push themselves to the edge of their abilities. No pain, no gain. Conversely, some pain can translate to a lot of gain.

I am not referring to physical pain, but rather the mental and sometimes emotional pain of making mistakes. Nor am I suggesting suffering as some sort of character-building exercise. The gain I'm describing here is about using errors as learning opportunities.

Error-driven learning is a well-established principle in the brain's motor system—that is, many neuroscientists believe that we learn to make skilled movements by observing the difference between what we intend to do and what we actually do. For instance, when musicians practice a song they already know fairly well, some parts will be relatively simple, but they might struggle to find the right chord in other parts. It would be extremely inefficient to record a new memory of every part of the song every time it is played. Instead, the better solution is to tweak the memory to better handle the challenging parts of the song.

Error-driven learning can also explain the benefits that come when we actively learn by *doing* rather than passively learn by *memorizing*. When you drive around a new neighborhood, you are going to be much better at learning the layout of the area than if you go through the same neighborhood as a passenger in a taxi. One of the many benefits of actively navigating in a new environment is that it gives you the opportunity to learn from the outcomes of your decisions and actions in a way that is not possible from simply looking at a map. Similar mechanisms are at play in numerous other activities. Whether you are an actor in a dress rehearsal, a football player running plays in a scrimmage, or a corporate executive practicing a presentation for your board of directors, you are intuitively exploiting the power of error-driven learning.

Memory researchers have long known that learning under challenging conditions has benefits. Consider the practical question of how students should prepare for a classroom test. The simplest way, used by most students everywhere, is to try to memorize the material by repeatedly reading the textbooks. Cognitive psychologists Henry "Roddy" Roediger and Jeff Karpicke considered a different approach: What if, instead of studying, you trained by *testing* yourself? Intuitively, it would seem like studying over and over is better than testing yourself—why risk producing wrong answers when you can just focus on memorizing the right one?

Roediger and Karpicke were willing to bet against the conventional wisdom because they had seen a number of studies that hinted at the power of tests as a learning tool. To examine the impact of testing, they had two groups of students memorize excerpts from a test-preparation book for the Test of English as a Foreign Language (TOEFL). One group memorized these passages by rereading them about fourteen times, and another group read the passages three or four times and then completed three tests in which they had to recall as much of what they studied as possible. The students who repeatedly studied the material were much more confident in their mastery of the material than the ones who were repeatedly tested, and their confidence was initially borne out. Those who repeatedly studied *learned* the passages a bit better than those who were tested, but that is not surprising because the study group had four times as many opportunities to read those passages as the ones who were tested. But then Roediger and Karpicke waited one week to see whether the students had made what they learned *stick*, and the differences were huge. On average, students who repeatedly studied retained only half of what they initially learned, but the ones who tested themselves retained over 85 percent. So, although students thought that they had learned much more of the material through studying, they actually got much more bang for their buck by testing themselves.

Roddy once wrote that, unlike in physics, there are no "laws" in the science of memory, but the benefit you gain from testing, as opposed to studying (aka the testing effect), is almost as reliable as the law of gravity. The testing effect has been shown in a massive number of research studies under a wide range of conditions. The effectiveness of testing remains undisputed, but scientists still don't agree on exactly why it has such a powerful effect on memory.

The simplest explanation is that testing exposes your weaknesses. In general, we tend to be overconfident about our ability to retain information we have just learned. The students in Roediger and Karpicke's experiment thought they were learning more by repeated studying because they had never been challenged. The ones who were tested had the humbling experience of struggling, and sometimes failing, to recall information that they thought they had learned well. As a result, it's possible that the students in the testing condition tried harder when they realized that they didn't learn as much as they thought.

But there's much more to testing than keeping us humble. Let's say you wanted to learn Swahili, but before you've even had a chance to study, you're asked, "What is the English translation for *usingizi*?" If you aren't a native speaker, you probably won't know the right answer, so you'll have to try your best to come up with a guess before you get the real answer: "sleep." This would seem like a terrible use of testing because struggling to answer a question on something you've not yet learned runs counter to how we think of education. Surprisingly, this kind of "pre-testing" turns out to be remarkably effective in learning. Why would it be a good thing to make your brain struggle to come up with potentially wrong answers before it even has a chance to learn the right answer?

Conventional wisdom would suggest that generating wrong information should be ineffective or even counterproductive because it leads to competition and interference, and most theories in neuroscience would predict that, if anything, it would be better just to give your brain the right information in the first place. Yet, something about giving your brain a chance to actively struggle seems to help us learn more and retain that information over time.

In 1992, cognitive psychologists Mark Carrier and Hal Pashler proposed an intriguing theory to explain this principle of memory. In computer science, it's well-known that machines learn a lot when they have to struggle. The neural network models that form the backbone of modern artificial intelligence systems learn through trial and error, by tweaking the connections between artificial neurons so that they get better and better at pulling up the right answer. Carrier and Pashler proposed that humans might also benefit from this sort of error-driven learning.

To see how this might happen in the human brain, Xiaonan Liu, a scientist in my lab (now a professor at the Chinese University of Hong Kong), my graduate student Alan Zheng, Randy O'Reilly, and I simulated the testing effect in our neural network model—the "hippocampus in a box." When we followed standard theories of memory in the brain and assumed that the hippocampus records everything that comes in, the model could memorize new information, but it was poor at retaining what it had learned. This is similar to what Roediger and Karpicke found in students who tried to learn from repeated study. When we changed the model to incorporate error-driven learning, we effectively turbocharged the hippocampus. It could learn and retain much more information in the face of interference.

When we looked under the hood of our hippocampus model, we could see that the benefits of testing do not come from making mistakes per se, but rather from challenging yourself to pull up what you have learned. To understand why, let's go back to our cell assemblies analogy. When you test yourself, your brain will try to generate the right answer, but the result isn't quite perfect. Your brain will struggle a bit and come up with a blurry approximation of what you learned. But this struggle provides a huge opportunity to learn. Stress testing your memory like this exposes the weaknesses in the cell assemblies so that the memory can be updated, strengthening the useful connections and pruning the ones that are getting in the way. Rather than relearning the same thing over and over, it's much more efficient to tune up the right neural connections and fix just those parts that we are struggling with. Memory updating is the key, because the most efficient way for our brains to save space and learn quickly is to focus on what we didn't already know.

Although we usually benefit from error-driven learning, there is one important boundary condition. Error-driven learning works if you eventually get close to the right answer, or at least if you can rule out wrong answers, so you have the opportunity to learn from your mistakes. You don't benefit from mistakes if you have no idea what you did wrong. You want to struggle, not flail around aimlessly. That's one reason it can be so hard to learn complex skills such as surfing; so much is going on, it's hard even to know what you've done right when you succeed and what you've done wrong when you fail. In these situations, expert instruction can be incredibly helpful to give you an idea about what you are supposed to be doing, as well as feedback specifically on what you have done right and where you went off the rails.

SPILLOVER

Memory is not a collection of isolated islands; it's an ecosystem of interacting cell assemblies. When we do anything to strengthen one cell assembly—such as repeatedly retrieving a particular memory—a ripple effect can extend to other memories occupying a similar place in the ecosystem. Sometimes, recalling one memory can make it harder to pull up related memories, a phenomenon called *retrieval-induced forgetting*. In the lab, you can show this by having someone study a bunch of words—say, the names of different kinds of tools such as *hammer* and *screwdriver*. Repeatedly testing someone on *hammer* will strengthen the memory for that word, but conversely weaken the person's memory for *screwdriver* if it was not tested. Recalling real events from one's life (aka autobiographical memories) can have a similar effect, whereby competing memories get inhibited—which might explain why my parents' recollections of my childhood tend to focus on a few memorable stories at the expense of many other events that didn't make the cut.

Fortunately, recalling something doesn't always have negative consequences for the stuff you do not remember. Under the right circumstances, recalling one piece of information can spill over to other memories in a positive way, strengthening both the memory you pulled up and related memories. In one study, volunteers were tested on facts learned in articles they read about toucans, the big bang, and (my favorite) Shaolin kung fu. When people were tested on one of the facts in an article, the benefits of that test spilled over to related facts, a phenomenon known as *retrieval-induced facilitation*.

Why would recalling some memories sometimes weaken and sometimes strengthen related memories? Xiaonan examined this in his computational model, and the answers fit with a lot of the principles we explored earlier. Let's start off with retrieval-induced forgetting. When disconnected memories overlap—say you ordered pizza at your favorite Italian restaurant a couple of weeks ago and then ordered lasagna when you visited the same place yesterday—you get competition because the cell assemblies that back different memories are constantly competing with one another in a Darwinian cage match. So, if you recall your lasagna dinner, you reinforce the cell assembly that backs the memory you just pulled up, and as a result the cell assemblies backing competing memories (e.g., your pizza dinner) will be relatively weaker and harder to pull up. Error-driven learning strengthens the memory you pulled up and simultaneously punishes the competition.

In Xiaonan's simulations, we also got a better insight into retrieval-induced facilitation. Suppose you splurged on a bottle of Brunello di Montalcino with your lasagna (and why not?). Because the lasagna and the wine were part of the same event, they don't compete with each other, so when you repeatedly recall that great lasagna, you'll incidentally also pull up your memory of that full-bodied red wine from Tuscany. The shared episodic memory means that an alliance is built between the cell assemblies that back the different components of the lasagna dinner, and when we recall the lasagna, error-driven learning tightens up the alliance, pulling up other connected elements from the same event.

Research on retrieval-induced forgetting and facilitation has a number of practical implications about how to translate principles of learning from the lab to the classroom. More often than not, tests are incomplete because it's usually not feasible to test on every piece of material covered in a class. This means testing can sometimes be counterproductive and result in retrieval-induced forgetting if learning and testing are not done in the right way. If students are encouraged to memorize a bunch of disconnected pieces of information—such as the dates of significant European battles in a history class—testing on some of those facts could heat up the competition and hurt retention of other facts that were not on the test. But if students are encouraged to develop a rich internal understanding of how the facts fit together—for instance, by understanding how those battles were connected to the larger political movements going on at the time—then testing can efficiently improve students' retention.

MAKE SPACE FOR LEARNING

Testing yourself isn't the only way to get the most out of error-driven learning. Error-driven learning doesn't require you to make a mistake—it happens whenever your brain struggles to pull out the right memory. The benefits that come about from stress testing the connections in your brain can be maximized not only by optimizing *how* you learn but also by optimizing *when* you learn.

Virtually all students have found themselves cramming for some exam. When I was in college, I pulled many all-nighters, waiting until the night before an exam to study several weeks of material. I found this to be exceptionally effective in the short run, but sadly, most of what I had learned would slip away within a few days after the end of the semester. Instead of studying six hours straight, the better way to learn and retain information is to break up your study time into several shorter sessions that are spaced out in time. A veritable mountain of findings in psychology show that you can generally get much more bang for your buck by putting gaps between your learning sessions rather than by spending the same amount of time cramming.

In 2021, James Antony, another scientist in my lab (now an assistant professor at Cal Poly), worked with Randy O'Reilly's hippocampus model to investigate why this happens. His simulations suggest that the benefit you get from spacing out your learning—known as the spacing effect—comes from error-driven learning. Take a moment to recall chapter 2 (see what I did there?), where we broke down how the hippocampus generates episodic memories by tying our experiences to a particular context—i.e., a specific place and time. It's not too hard to recall information you recently studied because your mental context has not changed so much since then. So, in the short term, cramming the night before for an exam can be good. But as time passes, the contextual state in your brain keeps changing, and it becomes harder and harder to get back to the information you learned back when you were jacked up on caffeine at 3:00 a.m. in your dorm room.

But what happens when you space out your learning? Suppose you read my chapter on context and episodic memory on the couch in your living room, then the next day you take this book with you to the beach and reread the chapter in a new context. At first, the hippocampus can pull out the memory of the last time you read the chapter, but it will struggle a little because you are seeing the same information you saw before, but in a different context. As a result, the hippocampal cell assemblies reorganize so that they place more emphasis on what you read, and the information is a little less tied to where and when you first read the chapter. James's modeling shows how, if you keep returning to that same information periodically, spacing out your study periods in between, the hippocampus can continually update those memories until they have no discernible context, making it easier to access them in any place at any time.

This goes back to the principle of "memory updating"—when we revisit the same memories over and over, they become repeatedly updated so that they lose those distinctive bits that can take us back to a unique moment in time. The upside is that those memories can become easier and easier to access, precisely because they are not tied to an idiosyncratic moment from the distant past. You don't have to go to the exact place where it happened or the mindset you were in at the time—the memory just comes up when you are looking for it.

TURNING MEMORIES INTO WISDOM

Error-driven learning is a fundamental principle that explains the dynamic nature of memory. I believe that it even reaches beyond our conscious experiences, extending into the myriad activities that go on in our brains while we are fast asleep.

That's right—our brains are hard at work during sleep.

Some of that work is housekeeping, such as clearing out metabolic waste that accumulates during the day (such as the beta-amyloid protein that is prevalent in the brains of patients with Alzheimer's disease). The brain also seems to take advantage of sleep to clean house with our memories, reorganizing them by seeking out connections between different events we have experienced.

The brain goes back and forth between at least five different states in a night's sleep. Levels of neuromodulators such as acetylcholine and hormones such as cortisol—the chemicals that help to stabilize the connections formed between neurons when we learn something new—change dramatically during these states. If you record the ongoing electrical activity in the brain (aka the electroencephalogram or EEG) during sleep, you can see that each stage has its own signature pattern of brain activity. There is no shortage of theories about how all these processes during sleep might affect our waking activities, including learning and memory, but the data has a lot of gaps, and we are still far from definitive answers. Nonetheless, the emerging evidence suggests that *slow-wave sleep* (SWS) and *rapid eye movement* (REM) sleep work hand in hand to transform our recent experiences into knowledge that we can use.

SWS, the deepest sleep stage, is the one that has been most reliably linked to learning and memory. When you are sleeping like a rock, your brain is working hard. EEG recordings reveal a wonderfully orchestrated interaction between the hippocampus and the neocortex during SWS. Large, slowly traveling electrical waves cycle across the neocortex, with smaller waves of activity called spindles riding atop the crest. Meanwhile, in the hippocampus, little bursts of activity called ripples bubble up, and during each ripple, individual neurons in the hippocampus that were active during the daytime come back to life, firing off in little sequences. Ripples, in turn, trigger bursts of activity in the default mode network (DMN), which helps us use schemas to learn about new events, and in the prefrontal cortex, which helps us to intelligently use schemas to form memories for events and reconstruct them later on.

In contrast to the intricately timed orchestra of neocortical and hippocampal rhythms that play out during SWS and REM, the neocortex starts playing free jazz. Researchers sometimes refer to REM as "paradoxical sleep" because the EEG during this stage resembles the fast bursts of activity you see when people are awake. But unlike when we are awake, most of us can't walk around during REM sleep, and we typically aren't seeing or hearing anything, so the neocortex is somewhat disconnected from the outside world. Dreams occur during REM sleep, and the dynamics of neocortical activity during REM might explain the vivid, lifelike experiences and bizarre logic that accompany our dreaming life. During REM sleep, the brain is generating its own sensory input and tries to make sense of it in the form of dreams, constructing an alternate reality all while you are resting safely in the confines of your bed.

Many in the neuroscience community have proposed that SWS plays a critical role in solidifying the memories of your daytime experiences, the process of memory consolidation. This idea is appealing in its simplicity, but it doesn't fully fit the data. The consolidation theory would predict that sleep is just like testing yourself. If memories are reactivated when we are tested, and memories are reactivated during sleep, they should have similar effects. Xiaonan Liu ran a number of studies testing this hypothesis and found that sleep doesn't *just* strengthen memories, it's much more interesting than that.

Consistent with some prior work, Xiaonan found that when we recall episodic memories during the day (e.g., in the context of a memory test), memories from competing events suffer from retrieval-induced forgetting, the phenomenon where the act of remembering one thing causes other related information to be forgotten. Xiaonan's research shows that sleep doesn't have the same effect. Instead, it seems to play a specific role in undoing the collateral damage that occurs when we retrieve particular events. After a night of sleep, people had better memory for events that were related to what they had previously been tested on, or retrieval-induced facilitation.

When we submitted these results for publication, we got a lot of flak from reviewers who couldn't understand how sleep could convert retrieval-induced forgetting into retrieval-induced facilitation. To get to the bottom of this, we used Xiaonan's computational model to understand the difference between what happens in the brain when memories are reactivated during a memory test taken while we are awake and what happens when memories are reactivated during sleep.

Xiaonan's simulations suggest that, in both cases, error-driven learning is the key. In our simulations of what happens when we are awake and recall a memory, error-driven learning improved the resilience of cell assemblies that carry episodic memories for the information that was recalled. However, because memory is an ecosystem, retrieving one memory can suppress the cell assemblies that carry competing memories learned in other contexts, resulting in retrieval-induced forgetting. In contrast, sleep may create an environment where cell assemblies active during different events play well together, rather than competing. When information from one event was reactivated in the model, the neocortex used this information to initiate a chain reaction of free associations. Information from different events was reconciled through error-driven learning, bringing out what was common across different events. The model suggests that when memories are reactivated during a test, error-driven learning helps to strengthen those specific memories, but when memories are reactivated during sleep, error-driven learning helps the brain to use threads of disparate experiences to weave a tapestry of knowledge.

Our modeling aligns with results from several studies suggesting that sleep helps us integrate what we have recently learned across different events so we can use that information more effectively. Words we memorized before sleep can more easily be incorporated in our everyday language after sleep. Memories for events sometimes become less context-dependent after sleep, such that we are able to more flexibly pull up information that we learned even when we were in a very different place. After sleep, we're sometimes better at seeing big-picture relationships between small pieces of information that we had previously learned, and we are better able to use this information to solve problems. To paraphrase sleep scientist Matt Walker, "Sleep can allow us to convert *memory* into *wisdom*."

The effects of sleep on memory make sense when we consider that the hippocampus and neocortex store different kinds of information. The hippocampus helps us pull up specific patterns of brain activity that take us back to a particular place and time, whereas the neocortex stores the semantic knowledge that enables us to understand what happens in an event and make predictions and inferences in new situations. During SWS, the hippocampus can ignite cell assemblies that captured important experiences from the previous day (episodic memory), and then during REM, the neocortex can play around with this information, free-associating to discover possible connections between different events we've experienced. The lesson to be learned is that an all-nighter is not all it's cracked up to be. Not only will we quickly forget what we have learned, we'll also deprive ourselves of the opportunity to get some rest and let our brain convert memories for events into knowledge we can use.

The memory benefits we gain from sleeping through the night might also be achievable during the day. Revolutionary research by UC Irvine psychologist Sara Mednick suggests that many similar benefits can emerge from a brief daytime nap. Even when we are awake, if we take a moment from our goal-focused activities to rest and zone out, we might accrue benefits that resemble what happens after we sleep. For instance, the ripples of neural activity that scientists see in the hippocampus of sleeping rats are also evident when a rat takes a brief respite after finishing a task. In humans resting in an MRI scanner, brain activity patterns change according to the task that they performed immediately beforehand—different areas of the brain that were engaged during the task continue to vigorously communicate with one another while we are resting, and this communication seems to help us retain what we have learned. In our lab, we have found that when we are resting, the hippocampus reactivates memories of contexts that were especially rewarding, helping us to prioritize memories for the things that matter.

Extrapolating to real life, after a challenging bout of work, it can help to get a good night's sleep, take a nap, or at least take a little time to rest. During these down states, our brains can use error-driven learning to piece together elements from different experiences, potentially allowing us to see things from a different perspective, giving us leverage to tackle problems that previously seemed insurmountable.

INCEPTION

When I was in graduate school, my adviser, Ken Paller, was considered an innovator in the field of memory research, and in the years since I graduated, he has become known as a pioneer in the field of sleep research. Ken was initially drawn to the field of neuroscience because he was fascinated with consciousness. By studying sleep, he has been able to bring together his interests in memory and consciousness. During sleep, we seem to be unconscious, yet the brain is hard at work tweaking our memories.

For a while, Ken viewed sleep research from the sidelines, thinking about what happens in the brain while we sleep. Then he had a crazy idea: What if you could hack the sleep process by triggering the reactivation of specific memories during sleep? I call this a crazy idea because, at the time, the standard thinking among researchers was that during sleep—and especially during SWS—you are unconscious, and therefore your brain is relatively cut off from the outside world. As a result, it should have been impossible to cue sleeping people to recall information. Ken, however, was impervious to the conventional wisdom and speculated that we might be able to hack into sleep to reactivate memories from specific events. Like the Leonardo DiCaprio character who infiltrates people's dreams in the film *Inception*, Ken wondered whether something similar could be executed in his lab—though with far less nefarious or unethical implications.

To test his unorthodox theory, Ken's team had people learn information while sounds were playing in the background—such as a picture of a cow accompanied by a moo sound. Next, the subjects took a nap, and some of those sounds were played again while they were in deep sleep. After waking up, the volunteers were unaware that the sounds had been played during the nap. Nonetheless, the sounds affected their memories for what they had learned from the previous day. The volunteers had more detailed memories of those study events (e.g., the exact location of the cow on the computer screen) that were reactivated while they slept.

Since then, Ken's team has explored numerous ways that this technique, called *targeted memory reactivation*, can be harnessed to improve memory and cognition. Ken's team has shown that targeted memory reactivation can improve learning a video game or a new language and can stimulate creative approaches to solving a problem that seemed unsolvable before sleep. They have even shown that reactivating memories that contradict common racial and gender stereotypes could reduce implicit bias. Ken is currently building on this work to explore just how far we can go to harness the power of sleep.

My UC Davis colleague Simona Ghetti has used a similar technique to probe memories in very young children. She had a group of toddlers (two to three years old) come to her lab on three occasions, where they got to play with a stuffed animal while a song played in the background. Within two days of their last visit to the lab, the children were brought to an MRI scanner in the evening, where they took a nap while their brains were being scanned. Although the toddlers were sound asleep, activity in the hippocampus spiked when Simona's team played songs the children had previously heard. The amount of activity in the hippocampus was directly related to how much information the child had learned about the stuffed animals during their play sessions. Simona's studies are providing an unprecedented window into the neurobiology of memory in toddlers, and they suggest that in babies, as in adults, the sleeping brain is highly receptive to the outside world.

RESHAPING LEARNING

Our emerging understanding of error-driven learning has important implications for educational practices. For generations, educators have seen testing as a way to separate the wheat from the chaff, and students have studied for tests by reading and rereading their textbooks. The idea is that, with each new exposure, they will pick up that much more information. But can you imagine a director of a play telling actors to prepare for opening night by simply reading the script over and over, without ever attempting to recite their lines from memory?

The implications of error-driven learning go well beyond helping students memorize information more effectively. It highlights the importance of adopting the right cultural attitudes about learning. Traditional approaches to education are about outcomes—did you learn the material or not? In China, India, and other Asian countries, standardized tests (such as the *gaokao*, China's notoriously grueling college entrance exam) are used as a yardstick of achievement, and they have massive effects on a student's future. In the United States, grades play a more important role than standardized tests (particularly for college, medical school, and law school admissions), but to get the best grades, you have to consistently perform well on tests. Unfortunately, these kinds of tests are optimized for last-minute learning that enhances short-term performance at the expense of retention.

We also seem to work against the brain's natural mechanisms for building knowledge. We start the school day early in the morning, making it difficult for students to get a good night's sleep, and we give them little time to rest or reflect. By discouraging good sleep habits and depriving them of downtime for restorative activities, students don't have the chance to take advantage of the brain's learning mechanisms that allow us to discover connections and convert what was learned into useful knowledge.

Growing up, I always felt I had one job to do in school, which was to perform well on all my tests. I now see that we have the potential to do so much more. Lasting benefits come from error-driven learning, during which we are not always going to be successful. We learn and retain more from the struggle of pushing ourselves to the edges of our knowledge than we do by memorizing and regurgitating on command. Perhaps instead of rewarding success, we should normalize mistakes and failures and incentivize constant improvement. Rather than emphasizing mastery, we need to celebrate the struggle—working to learn, rather than to prove you have learned something.

10

WHEN WE REMEMBER TOGETHER

HOW MEMORIES ARE SHAPED BY OUR SOCIAL INTERACTIONS.

. . .

> Our memory is made up of our individual
> memories and our collective memories.
> The two are intimately linked.
>
> —Haruki Murakami

We think of our memories as our own, but what we remember is inextricably intertwined with our social world. Humans are social animals, and our brains were undeniably shaped by evolutionary pressures to communicate and cooperate with those around us. Some scientists even argue that our capacity for language evolved primarily for the purpose of communicating memories to one another. According to one analysis, 40 percent of our conversational time is spent on storytelling and on creating and exchanging collective memories. Part of what makes humans unique is this existential pull to share our experiences with others to orient ourselves in the world. And memory plays a central role.

I became aware of the profound effect of social interactions on memory during my time at the Chicago Westside VA. One of my tasks as a psychology intern was to lead therapy groups for veterans. I was familiar with the scientific research that pointed to the efficacy of group therapy, but I wasn't confident about the endeavor. I expected it would be like a watered-down version of individual psychotherapy, and that I would end up having minimal interactions with the whole group at once. Most health insurance companies set limits on the number of sessions one can have, but the VA allowed its patients to drop in whenever they wanted, and the group had existed for many years before I joined. Many of the veterans were old-timers. Over the years, they had seen psychology interns come and go, but they knew one another's stories intimately. For these men and women, the VA hospital was a gathering place where they could connect with one another and collectively process their trauma. In contrast, I was a newcomer who had never been in the military, let alone seen combat.

I was apprehensive as I approached my first meeting with the group. I felt like an interloper who did not belong in the solemn space where these memories were shared. After I introduced myself, all eyes were on me. My clinical training had taught me that it's important to let a group member start off the conversation, but waiting for one of the veterans to break the silence was excruciating. The seconds felt like hours as I sat in suspense, waiting for someone to get things started. Finally, another newcomer to the group spoke up. He began by telling us his story of what had prompted him to seek help. After hearing his moving account and observing the way the other members leaned in to offer support and encouragement, I soon realized that my preconceptions about group therapy were completely wrong. Being in a group setting wasn't detracting from each individual's therapeutic experience; it was absolutely central to their treatment. And my role as an outsider was crucial for helping each participant process their memories and form bonds with other members.

My job was to guide the group, searching for common threads among their individual narratives of trauma and isolation and expressing those themes back to them. I was no longer just a therapist, I was now part of a team, collaborating with the group to create a deeper understanding of their experiences. We were using the power of the group to update people's individual memories and collectively construct shared memories centered on a new narrative that emphasized healing and empowerment.

Our personal memories do not exist in a vacuum—they are constantly being influenced and reconfigured as we interact with family, loved ones, friends, and our larger communities. The science of *collective memory*, a concept developed by French philosopher Maurice Halbwachs in the mid-twentieth century, is still nascent, but we have discovered that the very act of sharing our past experiences can significantly change what we remember and the meaning we derive from it. By exploring how social interactions and group dynamics affect the way we remember our lives, we are learning valuable lessons about how that influence can manifest in both positive and negative ways.

As we will see, the memories we construct and share as groups contribute to our sense of who we are. We construct our sense of identity, in part, through memories shared with our family and friends, as well as the memories shared across members from the same culture or nation. And our identities are built on ground that is constantly shifting, as we collaborate with one another to continually reconstruct and update our individual and collective memories. These memories are the lens through which we see our place in the world, as we come to terms with the past in order to make sense of the present.

THE STORY OF YOUR LIFE

Memories of our experiences and the meaning we derive from them are shaped by the people closest to us. Parents, peers, and partners all play a role in determining *what* we remember from our past and *how* we make sense of it. These complex social interactions come together to construct a *life narrative*—the stories we weave together to give us a sense of who we are (or think we are).

Developmental psychologist Robyn Fivush, whose research focuses on the social construction of autobiographical memory and narrative identity, has studied how everyday mother-child interactions can have a formative influence on how children remember their past. Let's say you take a three-year-old on a trip to the beach. You spend the day splashing in the waves, building sandcastles, and collecting seashells. One day later, if you ask the child to recall the outing, she might remember bits and pieces of the experience, but the way you interact with those memories will reshape her recollections, and her sense of self.

In her studies of mother-child reminiscence, Fivush found that children whose mothers asked open-ended questions ("What was the best thing we did during our beach trip?" or "Can you tell me what you saw when we walked around the rocks?") and elaborated on their children's answers ("After that wave knocked over your sandcastle, I loved how you went right back to building another one" or "You were so curious about those crabs in the tide pool") tend to remember more of their life experiences and put them together in a more coherent narrative than those whose mothers ask their children to recall specific information. These kinds of interactions can significantly impact a child's self-concept. Children who are encouraged to have a voice develop more ownership over their sense of self (e.g., *I am persistent; I am curious about nature*) because they are allowed to be authors of their personal narratives. Conversely, disallowing certain stories to be told or negating or disputing a child's perspective can undermine their sense of the experience and be detrimental to the development of the self.

When children remember an event during meaningful, engaging family interactions, it can have far-reaching effects on their sense of self as they get older. Fivush and her colleagues have shown that ten- to twelve-year-olds whose families discuss shared experiences collaboratively, knitting together their individual perspectives into a shared memory, tend to have higher self-esteem. Moreover, adolescents from families who discuss emotional aspects of their experiences are able to frame these recollections within a meaningful narrative, resulting in more confidence both socially and academically. Teenagers from families whose dinner-table conversations include shared reminiscences about family experiences from the distant past are less likely to be anxious or depressed and have fewer behavioral problems. These findings suggest that familial collaboration in the construction of autobiographical memories can play a positive role in children's development.

The dynamic interplay between the sharing of memories and one's sense of self is not limited to parent-child interactions. At any age, interactions with friends and family can reshape our memories in ways that affect how we see ourselves and, in some cases, lead us to fundamentally reshape the narrative. I experienced this firsthand some years ago during an ill-fated stand-up paddleboarding excursion with my longtime friend and UC Davis collaborator Randy O'Reilly, who, you may remember from chapter 8, designed the computer model that we use to simulate learning in the hippocampus.

Late one Saturday, just before sunset, Randy and I took what we expected to be a peaceful downstream paddle along Putah Creek, a deceptively placid waterway that runs through the city of Winters to the western edge of Davis. Soon after launching, however, we encountered a heavy current. My board hit something underwater and capsized. I surfaced, soaking wet and without my glasses, which had sunk at least twenty feet down to the bottom of the creek. Worse, the fins on the bottom of my board that would normally keep it from flipping over in the water broke off. Shivering and unable to see farther than ten feet in front of me, I began navigating the rapid current on the unstable board. We had no cell signal, and the steep banks were covered in thick foliage, so the only way out was to keep moving forward. Over the next hour, we faced one obstacle after another. I had to grab my board and clamber over enormous logs, duck under tree branches, and lie entirely flat on the board to slither under makeshift footbridges made of steel gratings with rusty bolts protruding barely an inch above my head. We even had to work our way around a generator that was hanging over the creek on a rope.

The situation went from miserable to terrifying when I got separated from Randy and the current tossed me off my board and wedged me between two fallen trees. I managed to cling to my board, but I'm not a strong swimmer, and I was hanging on for my life. I called out for help and could faintly hear Randy reply, but his words were drowned out by the thunder of the rushing water.

After several failed attempts to get back on my board, I realized that there was no point in panicking. No one was going to save me, and failure was not an option, so I'd have to calm myself in order to make it out alive. I took a few deep breaths, then managed to scramble back up on my board, pull myself out of the wedge, and reunite with Randy farther down the creek. The sun had gone down. Cold, wet, and exhausted, we spent another five hours paddling in the darkness through rapids, brambles, and thorny thickets before finally making it back to our car. With cell service restored, we each called our wives, who had been on the verge of mobilizing a search party.

Initially, I did not want to talk about the incident, but as my wife teased me about our colossally bad judgment, and my daughter commented on how lucky I was to be alive, my perspective on the experience began to shift. Later, as Randy and I told and retold the story to friends and colleagues, the tone gradually began to change. He and I would play off each other's memories of that day, weaving the ridiculous stream of obstructions and challenges we had to overcome into an epic tale comparable to the *Odyssey* (at least in our minds). Rather than focusing on my panic and incompetence, it became a story—and a memory—about digging deep to find the strength to overcome a significant obstacle. I learned that my general fear of what *could* happen is out of step with my ability to deal with things when they *do* happen. Through the supportive (and humorous) reactions to this story from my friends and relatives, I transformed a near-death experience into a new page in my life narrative.

The process by which storytelling can change our view of the past can be seen as an extension of the principles that govern individual memory. When we remember, our reconstructions of the past are dominated by the beliefs and perspective we adopt in the moment, in this case while we share those memories with others. We tailor narratives for our audience, who can reinterpret our memories and reflect them back to us from a different perspective. As we interact with the audience to reconstruct our past experiences, memories can be updated in the process, allowing us to see our personal past from a different view.

The transformative effect of social interactions on life narratives might be the "special sauce" that explains the efficacy of so many forms of psychotherapy, whether one on-one with a therapist or in group settings. In our brain imaging research, we have found that when people hear part of a story that can be stitched into a narrative with other story elements heard several minutes before, hippocampal activity increases, and the "memory code" in the hippocampus is transformed. It's not a stretch to think that collaborating with others to reframe and reinterpret your experiences can likewise lead you to discover new connections between different events in your past— connections that you can use to update your life narrative.

THE LOUDEST VOICE IN THE ROOM

You might expect that working with others to reconstruct an event will help us to remember more than if we were remembering alone. For instance, you would think that two people talking to each other about a baseball game they had recently watched should remember more plays than if they were each recalling the game on their own. However, in 1997, two research groups published papers that challenged this intuitive expectation. Across various experiments that required memorization of lists of words, sentences, or stories, the researchers compared the amount of information recalled by people working in groups against the amount recalled by the same number of individuals who had worked alone. Surprisingly, those working in a group had worse memory performance, a phenomenon called *collaborative inhibition*.

One reason for collaborative inhibition is that groups can magnify the effects of interference. Typically, when we revisit the past in a group, we have to wait our turn before contributing to the conversation. Meanwhile, information recalled by others can create competition in our recollections, leading us to forget details we might otherwise have remembered. If you watch a movie with a group of friends, then after one friend speaks at length about the poor performance of one of the actors, that might lead to retrieval-induced forgetting, such that you might struggle to recall moments that involved other actors in the film.

Recalling in groups can also have a homogenizing effect, filtering out the idiosyncrasies that make each of us remember things just a little bit differently. We tend to remember the information that we are moved to share with the others and leave out that which is unlikely to resonate with them, thereby morphing our memories in a way that makes them align with those of the rest of our group.

The selective nature of collective memory is not random—our memories become especially skewed toward those of the loudest voices in the room. In the lab, the recollections of groups disproportionately reflect the information recalled by those who dominate the conversation. These dominant narrators have an outsize influence by reinforcing the details they recall, which inhibits recollection of details from those who are less assertive. Collective memories are also dominated by the first person who speaks, and by those who are most confident in their memories of the shared experience, often at the expense of details that would have been contributed by those who don't speak as much or at all.

Suparna Rajaram, professor of psychology and cognitive science at Stony Brook University, has carefully studied how social interactions lead to the pruning of collective memory. This work shows that as people get repeated opportunities to recall information with group members, they begin to converge on the same memories. To see how this can play out in larger groups, she has used a technique called *agent-based modeling*. Just as neural network modeling simulates how neurons interact with one another to store and retrieve memories, agent-based modeling simulates how memories emerge and are transformed through interactions between people. Her results suggest that, as groups get larger, the whole is still less than the sum of its parts. This is because interactions between individuals reduce the diversity of memories across the group, leading their collective memories to become more and more homogenous with increasing numbers of interactions. The homogenizing effect of collaboration was even higher when these effects were simulated in social networks where group members started out with similar memories or beliefs, such as those with strong familial and social bonds or a common national, religious, or cultural identity.

Fortunately, collaborating doesn't always inhibit individual memory. We have seen how we can minimize interference (i.e., competition between memories) when we focus on distinctive elements of particular experiences, and this principle applies to groups as well. When members of a group collaborate closely, making sure to consider the unique contributions of each, the collective memory of the group is often greater than the sum of its parts. *Collaborative facilitation*, as this is known, tends to happen when people have shared expertise that enables them to work as a team.

Facilitation can also occur when people have a close relationship with each other. For instance, psychologist Celia Harris at Macquarie University found that the collaborative dynamics between couples can facilitate memory performance when they work together to recall information. One factor identified in her research is that long-term couples are often able to generate the cues that help each other remember better. This is true for simple stimuli such as word lists, and similar dynamics also appear to be at play when intimate couples interact with each other to recall events that they experienced together.

Not all couples benefit from collaboration—some remember less when they work together. The key to successful collaborative memory seems to lie in having some common ground, along with an appreciation for each other's distinctive contributions. Harris reported one such exchange from a couple who effectively cued each other when asked to recall details from their honeymoon forty years earlier:

> WIFE: And we went to two shows—can you remember
>   what they were called?
> HUSBAND: We did. One was a musical, or were they both?
>   I don't . . . no . . . one . . .
> W: John Hanson was in it.
> H: *Desert Song.*
> W: *Desert Song*, that's it, I couldn't remember what
>   it was called, but, yes, I knew John Hanson was
>   in it.

Some findings even suggest that as couples age together, they become more cognitively interdependent and share a similar trajectory of mental abilities. One particularly promising line of research in this area has focused on the effects of aging on cognition in older couples. It's well established that, although our brains change as we age, these changes are not necessarily accompanied by declines in memory and other cognitive abilities. Working with others to share the load of remembering can provide a critical aid to compensate for the effects of brain damage. For instance, people with memory disorders can handle challenging problem-solving tasks through interacting and finding common ground with their spouses. Although more research needs to be done, there is good reason to think that older adults suffering cognitive decline may function at high levels when they can lean on their partner or spouse.

Members of a couple don't have to use the same memory strategies or focus on the same information. As long as they're working together, the benefits can be even bigger if each partner remembers events in a different way. My wife is more likely to remember the conversation topics at a dinner, whereas I'm more likely to remember what we ate and how much we paid, but when we reminisce together, we recognize each other's idiosyncrasies and integrate our stories into a shared narrative. Collaborative facilitation works when people understand and value each other's expertise and actively incorporate each perspective into a shared narrative.

SOCIAL DISTORTION

We've seen how, as we remember the same event repeatedly, a memory becomes like a copy of a copy of a copy, increasingly blurry and susceptible to distortions. Frederic Bartlett demonstrated this effect in individuals nearly a century ago with his experiments on remembering. When his University of Cambridge volunteers repeatedly recalled a Native American folktale, each iteration became increasingly distorted as they recalled the story through the schemas of their own cultural expectations and norms. Having studied cultural anthropology, Bartlett was curious how these distortions might play out when we transmit information to others.

To address this question, he ran an experiment on "serial reproduction," in which his Cambridge student volunteers were shown a simple line drawing of an African shield, then asked to redraw it from memory. Bartlett then gave the students' drawings to another group of volunteers and asked them to reproduce the images from memory; these drawings were then reproduced from memory by a third group, and so on. As memories of the image were passed on from person to person, the drawings gradually began to look less like an African shield and more like a man's face. The sharing of memories for the shield morphed it into something the volunteers were familiar with and could easily communicate given their shared cultural knowledge.

Memory researchers have since used Bartlett's serial reproduction method to investigate how memories are distorted as they are passed within groups. In one experiment, volunteers were asked to memorize a story that included some information that was consistent with commonly held gender and social stereotypes (e.g., a football player drinks beer with his mates as they drive up the coast to a beach party) and some that was inconsistent (the footballer switches the radio station to classical music and stops at a roadside stand to buy himself flowers). When volunteers heard the original story, they recalled both stereotype-consistent and -inconsistent information. By the second time the story was passed on to a new listener, the stereotype-consistent information remained, but the inconsistent information was lost. The information degraded even further from person to person if the information was passed on to people who endorsed the stereotype embedded in the story. This suggests that when we share memories of an event, information that conforms to our stereotypes is passed on to others, whereas inconsistent information is more likely to be lost. As a result, our collective memories can reflect and reinforce our preconceived ideas and biases.

Other findings suggest that we also have a negativity bias in social transmission of memories—as memories are shared from person to person, negative information (such as a story about corrupt behavior by a politician) is more likely to survive, whereas positive information (such as when a politician writes a bill to reduce corruption) is more likely to be lost. Even when events that are ambiguous are shared from person to person, they are more likely to be passed on with a negative spin.

Distortions in collective memory are not just the result of preexisting biases. The mechanisms of memory are imperfect, and when information is communicated within a group, errors start to accumulate. Cognitive psychologist Roddy Roediger, who has extensively studied memory distortion in individuals, coined the term *social contagion* to describe how memory distortions seem to spread like viruses through social networks. To understand how this works, Roediger came up with an experiment in which groups of two people were given a set of photos to study, then asked to recall the items they remembered from those photos. There was, however, a catch:

only one member of each "group" participating in the experiment was an actual volunteer who faithfully attempted to remember what was studied; the other member, who was planted by Roediger, deliberately recalled a few objects that were never shown. As you might expect, the true volunteers became "infected" by the misinformation and were more likely to recall the items that were previously recalled by the duplicitous collaborator.

One interesting aspect of Roediger's experiment focused on attempts to either inoculate his volunteers from social contagion or to root out the false memories implanted by the collaborator. Even when the volunteers were warned that their collaborator might have mistakenly recalled items that were not in the set, they were still likely to remember the misinformation planted by their partner.

When we share memories in groups, some people are especially effective at spreading misinformation. Those who dominate a group conversation are more likely to spread their memory errors to the rest of the group, as are those who are more confident or speak first. Sadly, we are even more susceptible to inheriting memory errors when they come from friends or others we trust. All these findings point to one of the most insidious consequences of collective memory: once distortions creep into our shared narratives, they can be incredibly difficult to root out.

There are some factors that can protect us from social contagion. For instance, we tend to be more resistant to social contagion when we get information from people who are not perceived as credible sources (although, unfortunately, this tends to be children, the elderly, or individuals from an out-group). On a more positive note, we are also less susceptible to collective memory distortions when information is passed on interactively—when we actively engage with others, we are less likely to incorporate misinformation than when we passively receive information, say through social media.

The tendency of groups to magnify and transmit memory errors can easily be exploited to spread misinformation, as evidenced by the viral spread of false information about the 2016 Brexit referendum in the United Kingdom, baseless election-fraud claims in the 2020 U.S. presidential election, and conspiracy theories about the COVID-19 vaccine. The rise of misinformation, planted in the public sphere as "news," has even led to a new branch of applied research focused on the psychology of "fake news."

Long before the term *fake news* became a standard entry in English dictionaries, Elizabeth Loftus and colleagues studied how the mechanisms for implanting a false memory in an individual could be weaponized to spread false information throughout a population. In collaboration with *Slate*, an online magazine that covers politics and current affairs, Loftus's team ran an online experiment in 2010 to test people's susceptibility to forming memories of fabricated news. The study, with over five thousand surveyed, used a watered-down version of the memory-implantation recipe Loftus had developed in her lab nearly two decades earlier: Volunteers were exposed to a series of accurate news stories accompanied by actual photographs, and a fake news story accompanied by doctored photographs. One fabricated story read, "President Obama, greeting heads of state at a United Nations conference, shakes the hand of Iranian president Mahmoud Ahmadinejad." The story was accompanied by a photo-shopped picture of Barack Obama shaking Ahmadinejad's hand. In real life, the two never met in person. Nonetheless, almost half of the participants in *Slate*'s informal experiment who were exposed to the photo said that they remembered having seen that article in the news one year earlier. One *Slate* reader commented, "The *Chicago Trib* had a big picture of this meeting," and another said, "I remember most the political hay Republican bloggers made about the handshake." Overall, up to a third of the respondents inaccurately recalled the fake news incidents as news stories that they had previously read about, and when forced to choose, many were unable to tell real stories from the fabricated ones.

What makes people susceptible to fake news? Part of what makes the human brain vulnerable to social contagion comes from our bias to believe and therefore remember information that is consistent with our preexisting beliefs. Fake news is easier to digest if it comes in a flavor we already like. Consistent with research on social contagion, belief in fake news is also increased when the information is emotionally arousing, when it includes photos as well as text, and when it comes from a source we know and trust.

Fluency also plays an important role in social contagion. If we are repeatedly exposed to false information, it can become increasingly familiar, influencing our perception of the truth. As we have seen, we can gradually learn semantic knowledge through repeated exposure, and similar mechanisms can lead us to acquire beliefs that are not based in fact. We have all heard widely believed "facts" that are actually myths—*the Great Wall of China is visible from space; the MMR vaccine causes autism; Vikings wore helmets with horns on them; you only use 10 percent of your brain*. Those myths take on a life of their own because they are widely repeated. Social contagion might take place more rapidly if that information is repeatedly shared by different people in our often-insular social media networks.

Media consultants and political campaign strategists have mastered the art of spreading misinformation through social media channels, and memory researchers are still trying to catch up. Recent research has highlighted the use of "push polls," designed not to collect opinions but to spread misinformation. Push polls first rose to the public consciousness during the 2000 South Carolina Republican Party presidential primary between George W. Bush and John McCain. In the run-up to the election, South Carolina residents were barraged with poll questions from the Bush campaign, such as "Would you be more likely or less likely to vote for John McCain for president if you knew he had fathered an illegitimate black child?" McCain had actually adopted a child from an orphanage in Bangladesh, so the question was a blatant and carefully crafted work of misinformation. Other questions planted insinuations that McCain had "proposed the largest tax increase in United States history" and wanted to "give labor unions and the media a bigger influence on the outcome of elections." When McCain learned of the push polls, he denounced the tactic, but the damage was done, and he lost the primary.

Push polls seem to work by letting misinformation burrow into our memory. In an experiment involving over a thousand participants in Ireland, researchers demonstrated the shocking extent to which they can contaminate an individual's memory. When recalling an innocuous story about a fictitious politician named Catherine, over half of the participants in the study incorporated misinformation provided in a push poll. The study further demonstrated how push polls could reinforce beliefs in, and memory for, fake news articles about real public figures such as Pope Francis, Irish prime minister Leo Varadkar, and Donald Trump.

Participants in research experiments like these are all exposed to the same misinformation, but in the real world, different social groups often rely on disparate sources for the information they consume and share. As we become increasingly segregated along cultural, racial, and political lines, social contagion can lead us to develop entirely different memories of the same events, and hence different views of reality. All this can lead to polarization and tribalism, whereby negative misinformation about "the other side" can take hold and exacerbate harmful stereotypes.

Fortunately, the effects of fake news can be mitigated by fact-checking, though the corrections need to be delivered in the right way. In one study, over five thousand participants read through head-lines with accompanying photos, some depicting actual events, while others were fabricated. The experiment was designed to see whether information from a third-party fact-checking website could be used to inoculate people against remembering the fabricated headlines as real events. Timing turned out to be everything. Fact-check warnings before or during readings had only a small effect on people's ten-dency to recall the false articles as true, but warning people afterward made them 25 percent less likely to endorse the false headlines. These findings suggest that seeking out a fact-check after consuming fake news can enable us to update our memories and thereby curtail the effects of misinformation.

BUILDING BETTER NARRATIVES

Although collective memory is currently a blip on the radar of most memory researchers, it is a major topic of interest among sociologists and historians, who view memory as a mechanism for the construc-tion of historical narratives that contribute to a sense of cultural and national identity.

Putting the science together, what we see is that the quirks of human memory create a perfect set of conditions that can give rise to groups constructing a selective and sometimes biased perspective on the past. In homogenous groups, we tend to home in on a particular perspective, selectively prioritizing information that is emotionally charged, consistent with our prior beliefs, and endorsed by people in power. This effect tends to be amplified in social networks, such that people who run in the same circles have more homogenous memories of the same experience. Social networks can also accelerate the spread of false information, due to both individual memory distortions and the sharing of misinformation from external sources.

The selectivity and malleability of collective memory can be readily demonstrated in random groups of strangers who have been brought together in a lab to remember things that are fairly arbitrary. But in the real world, as we collaboratively build the historical narratives of a culture or a nation, the stakes are even higher. The dynamics at play mean that those who have the power and privilege to speak first, speak the most, and speak most confidently will shape the collective narrative.

The positive lesson to be taken from research on collective memory is that we can benefit from differing perspectives. Just as collective memory suffers from homogeneity, it can also benefit from diversity. When we become aware of, or try to adopt, an alternative perspective, we become attuned to information that does not fit with our prior beliefs, giving an opportunity to correct misconceptions and overcome biases.

Collective memory research has shown that we remember more when group members actively seek to include input from each member, whether a group of strangers in a memory lab or partners in an intimate relationship. As a society, we can gain a better appreciation of our past by looking beyond wars and presidencies to consider the experiences and perspectives of everyday people, including those from marginalized communities. By learning from, and finding common ground with, a diverse range of individuals, we benefit from a variety of viewpoints that would otherwise be lost to competition in the historical "marketplace of ideas."

CODA

DYNAMIC MEMORIES

. . .

> Everyone is the poet of their memories. . . .
> But like the best poems, they're also never really
> finished because they gain new meaning as time
> reveals them in different lights.
>
> —Richard Hell

We began this book with a question: *Why do we remember?* Almost thirty years after pasting those sticky electrodes on the head of my first experimental subject, I still don't have a simple answer.

I'm okay with that.

In my circuitous journey through the world of learning and memory, I have learned that there isn't one answer to why we remember, because there isn't a single mechanism or principle that explains all the ways we are changed by and can call upon the past. Memory is the product of a brain that emerged over millions of years from a series of compromises—design choices imposed by the constraints of evolution, which come with both costs and benefits.

The myriad ways we use memory are enabled by interactions between evolutionarily ancient structures of the brain (such as the hippocampus, amygdala, and nucleus accumbens) and relative newcomers (such as the perirhinal and prefrontal cortex and the DMN), along with neuromodulators (such as dopamine and noradrenaline), which drive plasticity. The neural tweaks that enable us to fluently process what is familiar; the role of memory in orienting us to what is different, new, or unexpected; the use of episodic memory to predict what can happen in the future; the way we can update memories based on new information and learn from our mistakes; the ways in which the sharing of memories enables us to constantly reshape the narratives of our lives and the cultures that we are a part of—all these processes reflect the workings of an intertwined system, the precise nature and functions of which neuroscientists such as me spend our days and nights trying to understand.

I chose to name my lab at UC Davis the Dynamic Memory Lab in part because I think of memory as the process by which our brains change over time. As we go about our lives, connections between neurons are constantly formed and modified, resulting in cell assemblies that help us sense, interact with, and understand the world around us. These intricately connected neural networks give us the ability to weave together the threads of the past so that we may envision how the future will unfold.

As we progress from birth through old age, we can see that memory has a distinct role to play in each life stage. In most species an animal's life span primarily encompasses the period of time when it's capable of reproduction. Humans are relatively distinct in that our lives are typically bookended by a prolonged period of brain development prior to puberty and a lengthy life after fertility is no longer possible. It might seem odd that evolution would favor the survival of a species in which individuals spend most of their time incapable of reproduction. But it is possible that changes in our ability to learn and remember across the life span might have conferred benefits that were critical for our flourishing as a species.

Developmental psychologist Alison Gopnik has argued that the extended window of time that it takes for executive function to peak might indicate that our long childhood is made for a different mode of learning that is adaptive for families and society. She likens children to scientists, gathering information about the world through play and exploration. Their brains are optimized toward this kind of learning because they aren't overly constrained by any single goal, whereas an adult with a fully developed prefrontal cortex can care for children's needs by focusing on goals that are critical for survival and success. As a result, young people may be in a better position to discover what is new and see possibilities that adults, consumed by the day-to-day tasks of life, might miss.

We have also seen that episodic memory declines as we get older, leading us to increasingly experience the frustration of misplaced keys, forgotten names, and baffled moments when we forget what we were just talking about. But not all memory functions change with age—semantic memory, in particular, remains robust long into old age. The persistence of semantic memory in the face of episodic memory declines may be a clue about the functions of memory in old age, when life is not so much about new learning and chasing our own goals but about sharing what we've learned with the people around us. We can see the importance of this stage of life in indigenous cultures, where elders play a central role in the community by transmitting cultural knowledge about language, medicine, foraging, and hunting to younger generations.

Recent research suggests this role might not be unique to humans. In orcas (aka killer whales), another species in which the life span extends well beyond the period of fertility, transmission of cultural knowledge and traditions (hunting preferences, play behavior, even mate preferences) appears to be driven by postmenopausal females. As with human elders, the knowledge and experience accumulated over a lifetime by orca grandmothers plays a critical role in the survival of the pod.

Rather than seeing our lives as a single trajectory from maturation to eventual senescence and decline, we can view our lives as a series of stages. At each stage, memory is doing exactly what it has evolved to do, and in the process, it connects us with one another, grounding us in the social world.

As I have come to understand the role of memory in our lives, I am increasingly aware of the challenging task ahead. To fully understand why we remember, we scientists will need to find bridges between the temporal scales from milliseconds to hours to decades and spatial scales from the level of ion channels in a single neuron to vast networks of connected neurons to the social networks of interacting humans. It's a daunting task, one I doubt we can complete in my lifetime.

Fortunately, I'm not in it for the answers. Science is not about having all the answers; it's about asking better and more revealing questions. There is always going to be a missing piece of the puzzle. But searching for an answer forces us to see the world in new ways, challenging our most stubborn assumptions about who we are.

When we treat memory as if it should be a literal record of the past, we end up with unrealistic expectations and are left feeling perpetually frustrated. We remain prisoners of an incomplete past, oblivious to the ways in which the past has shaped our understanding of the present and our future decisions. Only when we start to peek behind the veil of the "remembering self" do we get a glimpse of the pervasive role memory plays in every aspect of the human experience and recognize it as a powerful force that can shape everything from our perceptions of reality to the choices and plans we make, to the people we interact with, and even to our identity. When we get to know the remembering self, we can seize the opportunity to play an active role in our remembering, freeing ourselves from the shackles of the past, and instead using the past to guide us toward a better future.

ACKNOWLEDGMENTS

It might seem odd, but in the academic world, most scientists don't get much respect for writing books. To land a faculty position, secure research funding, get promoted, and earn the respect of your colleagues, you have to publish lots of peer-reviewed academic research papers. Books generally don't factor into the equation. Consequently, my motivation to write this book was more personal than professional—perhaps some combination of altruism, activism, masochism, and narcissism.

This labor of love was written in the "spare time" left after teaching, dealing with the financial and bureaucratic responsibilities of running a large research lab, writing progress reports, faculty meetings, student advising committees, meetings with university leaders and donors, service on departmental, university, and professional society committees, peer review obligations, and, of course, actual scientific research. All this was further complicated by a global pandemic. There is no way I could have surmounted these obstacles on my own. Countless collaborators, colleagues, friends, and family members made this journey possible.

My agent, Rachel Neumann, and her partners Doug Abrams and Lara Love at Idea Architects saw the potential for me to write a "life-changing" book. Rachel, who is an accomplished writer herself, tirelessly advocated and negotiated on my behalf, while providing insightful feedback on each draft and continually pushing me to strive for perfection. Idea Architects' international partners at Abner Stein and the Marsh Agency connected me with publishers around the world. Jason Buchholz helped immensely on my book proposal. And the entire team at Idea Architects, including Boo Prince, Ty Love, Sarah Rainone, and Alyssa Knickerbocker, gave me the confidence to navigate the ins and outs of the writing and publication process.

I am lucky to be working with incredible publishing partners. At Doubleday: vice president, editorial director of nonfiction, and editor of three Pulitzer Prize–winning books Kris Puopolo (who also came up with the title for this book) and assistant editor Ana Espinoza; and at Faber & Faber: editorial director Laura Hassan and publishing director Hannah Knowles. They totally understood my vision for the book and helped me bring it to fruition.

I am especially indebted to Wenonah Hoye. From the beginning, she had an uncanny understanding of what I wanted to say and how to say it with the right balance of science and storytelling. Her optimism, tenacity, inquisitiveness, patience, and wisdom made it a joy to write this book.

A number of talented scientist-authors helped to get this project going. Amishi Jha (cognitive neuroscientist at the University of Miami and author of *Peak Mind*) played a pivotal role in prompting me to write this book, and Ethan Kross (cognitive neuroscientist at the University of Michigan and author of *Chatter*) spent hours on the phone educating me about the writing process. David Eagleman, Andre Fenton, Mike Gazzaniga, Joe LeDoux, Dan Levitin, Lisa Miller, Edvard and May-Britt Moser, Siddhartha Mukherjee, Tali Sharot, Robert Sapolsky, and Matt Walker took the time to read my book proposal and vouch for this first-time author.

I am grateful to the National Institutes of Health, U.S. Department of Defense, Office of Naval Research, James S. McDonnell Foundation, Leverhulme Trust, and the Institute for Research on Pathological Gambling and Related Disorders for supporting my research. And I was incredibly lucky to receive a Guggenheim Fellowship to support work on this book.

No one makes it far in this business without good mentors, and I had some of the best. Jerry Sweet at Evanston Hospital taught me the art and science of clinical neuropsychology, and he even came to see my band play a few times. I also thank the clinical internship staff at the Chicago Westside VA hospital and the veterans who trusted me and taught me so much. And my research career would not have happened were it not for my graduate school mentor Ken Paller and my postdoctoral mentors Mark D'Esposito and Marcia Johnson. I cannot put into words how much I learned from them. They continue to support me to this day, including reading and suggesting revisions on several sections of this book.

I've been fortunate to have wonderful colleagues read and provide feedback about various sections of this book, including James Antony, Felipe De Brigard, Elizabeth Loftus, Mara Mather, Chandan Narayan (my cousin and former bandmate, who is a psycholinguist at York University), Nigel Pedersen, and Suparna Rajaram. I am also indebted to Howard Eichenbaum, who took me under his wing early in my career. He was a formative influence on my scientific development, and I will never forget the example he set through his creativity, kindness, and ability to see what is special in others.

UC Davis is one of the best places in the world to study learning, memory, and plasticity. I thank Ken Burtis and Kim McAllister for making it possible for me to start the Memory and Plasticity (MAP) Program, and Dan Ragland for helping to bring my research back into the clinic. Andy Yonelinas, my close friend and unofficial guru for over twenty years, taught me about the complexities of human memory, inspired me through our collaborations, and helped me become a better scientist. He also tried to teach me snowboarding, though I haven't gotten very far on the error-driven learning curve. Discussions with Andy, Brian Wiltgen, and Arne Ekstrom at our weekly memory meetings expanded my thinking and encouraged me to challenge the status quo. My dear friend Simona Ghetti is one of the world's leading experts in the development of human memory, as well as memory accuracy and metacognition. She took the time to read and comment on the entire book, and she provided sage advice, both professional and personal, over the course of many dog walks. Randy O'Reilly taught me about neural networks, and in the process, completely transformed my thinking about how and why we remember. He and Yuko Munakata (who has a fantastic book on child development in the works) spent many hours talking with me about book stuff during paddleboarding and wake-surfing expeditions.

The foundations of this book were established over many years of teaching and mentoring UC Davis undergraduates. Our diverse student body, approximately 40 percent of whom are the first in their family to get a college degree, are as brilliant and enthusiastic as any students you'll find at the most prestigious Ivy League universities. The undergraduate students who volunteered to do research in my lab contributed to numerous studies that we have published. And I could not imagine getting any research done without the executive functions of my lab managers. Most of all, I am grateful to the graduate students and postdocs I have mentored or co-mentored, who continually open my eyes to the fascinating and counterintuitive aspects of human memory. Their contributions are embedded in the DNA of this book.

My parents, Sampath and Anu Ranganath, read entire drafts of the book and helped me improve the storytelling. More important, from my childhood into the present, they continually encouraged me, guided by the belief that I could do something extraordinary. I am eternally grateful to them, along with my brother, Ravi, and his family (Tiff, Charlotte, and Alice), my grandmothers, Shamala Ranga and Vijaya Sampath, and my second family, Charles, Laurie, and Kevin Ryavec, for their support and encouragement.

I continually draw inspiration from my daughter, Mira. I am grateful for her patience in hearing me alternate between despair and elation as I talked endlessly about the writing process. She is an intelligent, kind, and creative young woman, and it has been wonderful to see her develop her own scientific career in the field of plant biology.

My greatest debt goes to my wife, Nicole, without whom I'd have never embarked on the journey to become a scientist, let alone write this book. Her determination, curiosity, ingenuity, and resourcefulness inspired me on numerous occasions to overcome my inertia and to follow through on my instincts and create opportunities where none seemed to exist. She read every draft of the book and offered insightful feedback, and her encouragement and love kept me going through many hours of writing, hand-wringing, and kvetching. She did all this while working on her own book, which will restore the collective memory of those whose voices have been ignored. Our life together has been truly memorable.

NOTES

It is challenging to present scientific research in an accessible way because every take-home message requires an inferential leap from the lab to the real world. I have endeavored to make reasonable conclusions about the real-world implications of scientific research and to clearly distinguish between speculation and scientifically backed conclusions. For the sake of clarity, I avoided getting into the weeds and instead cover some of the nuances, caveats, and alternative interpretations of the available evidence in the annotations below.

This isn't meant to be an exhaustive list of references. To provide context, the literature on the neuroscience of memory is vast. In April 2023, a PubMed search on the term *memory* revealed 383,040 results, and this is surely an incomplete list. Even within the range of topics discussed here, I could not discuss or cite every relevant paper. The cited references were chosen in an effort to balance scholarship with provision of accessible entry points to the literature for interested readers.

INTRODUCTION: MEET YOUR REMEMBERING SELF

3 your "remembering self" makes the choices: See Kahneman and Riis 2005. Kahneman's masterfully written trade book, *Thinking, Fast and Slow* (2011), elaborates on the contrast between the remembering self and the experiencing self. The book is ostensibly about decision-making, but much

of his life's work has really been about how memory influences decision-making. As an aside, the upper-division class on decision-making that he taught when I was an undergraduate at Berkeley had an enormous impact on my thinking about human cognition.

4 "literal recall is extraordinarily unimportant": Bartlett 1932.

5 "Russia Under the Mongolian Yoke": Ian Gotlib, my adviser at the time, was the first author of the publication of this work (Gotlib, Ranganath, and Rosenfeld 1998). Also, in case you're concerned about the effects of our mood-induction procedure, don't worry—we played a tape of Dixieland jazz music to cheer up our subjects at the end of the experiment. The selections were so bad that you could not help but laugh.

6 Human memory needed to be flexible: Schacter (2002) makes this point nicely.

1. WHERE IS MY MIND?

11 average American is exposed to thirty-four gigabytes: Nick Bilton, "The American Diet: 34 Gigabytes a Day," New York Times, December 9, 2009, https://archive.nytimes.com/bits.blogs.nytimes.com/2009/12/09/the-american-diet-34-gigabytes-a-day/; Harris Andrea, "The Human Brain Is Loaded Daily with 34 GB of Information," Tech 21 Century, December 2016, https://www.tech21century.com/the-human-brain-is-loaded-daily-with-34-gb-of-information/.

13 The scientific study of memory: My description did not go into the details of Ebbinghaus's (1885) meticulous and somewhat counterintuitive procedure, known as the "method of savings." Unlike most modern memory studies, Ebbinghaus did not quantify the number of trigrams that he was able to recall after a delay. Instead, he must have reasoned that, even if he could not bring one of those studied trigrams to mind immediately, it might still be somewhere in his head. To get around this problem, Ebbinghaus would repeatedly study a list of trigrams until he memorized all of them. After some time, he worked on re-memorizing the list. Ebbinghaus assumed that if he had retained some memory for the trigrams that he had memorized, then he should be able to learn them faster the second time around. So, Ebbinghaus tracked forgetting by calculating the difference in learning time from the first to the second test, a measure he called "savings." These days, researchers rarely use Ebbinghaus's method of savings, but it doesn't seem to matter because the forgetting curves quantified by Ebbinghaus generally hold true even if you test memory in simpler ways.

If you're curious about Ebbinghaus but are not a glutton for punishment,

Henry "Roddy" Roediger (1985) wrote an easy-to-read synopsis of Ebbinghaus's book.

14 "exhaustion, headache, and other symptoms were often felt": Ebbinghaus 1964, chap. 6, "Retention as a Function of Number Repetitions," section 23, "The Tests and Their Results."

14 Although there are some caveats: Ebbinghaus mostly studied memory for relatively meaningless information (with the exception of an experiment in which he memorized the poem *Don Juan* by Lord Byron). It's less clear how well the Ebbinghaus forgetting curve applies to more lifelike events. For instance, you might remember a disastrous job interview from the distant past but forget many of the relevant details about what was said and who said it. See Radvansky et al. 2022.

14 massive populations of *neurons*: Herculano-Houzel 2012.

16 the science has been a bit distorted: Eagleman (2020) provides a science-backed perspective on neural plasticity written for a broad audience.

17 interference is the culprit: MacLeod (in press) provides a great review of interference theory. In my description, I've glossed over the debate over whether forgetting occurs because memories are erased or because we just can't find them. This is an important and interesting issue, but no one would debate that competition is a source of forgetting. Also, I have taken some liberties, collapsing across the many flavors of interference documented in the literature, because I think the main point is the same: in memory, as in life, the competition can be fierce.

17 When we face an onslaught: For more on this topic, along with practical tips to better manage your attention, see Gazzaley and Rosen 2016 and Jha 2021.

20 caused deficits in thinking and learning: Titles cited here are from Teuber 1964, Nauta 1971, and Goldman-Rakic 1984.

20 The prefrontal cortex was seen as a "working memory": I'm referring here to a debate during the late nineties and early aughts among neuroimaging researchers about whether the prefrontal cortex supports executive function or working memory maintenance. This latter view is usually attributed to Jacobsen (1936), and later Goldman-Rakic (1987). Many others in the field emphasized the role of the prefrontal cortex in short-term memory or working memory maintenance, but I don't think that Goldman-Rakic or any other experts on the prefrontal cortex believed that this was its *only* function. Goldman-Rakic's references to working memory were inspired by parallels between her ideas about prefrontal function and the psychological model of working memory proposed by Baddeley and Hitch (1974).

Broader treatments of the relationship between working memory and long-term memory are given by Wagner (1999), Ranganath and Blumenfeld (2005), and Jonides et al. (2008).

21   To study working memory with fMRI: See, for example, Braver et al. 1997 and Cohen et al. 1997.

22   there were the people with damage to the prefrontal cortex: Throughout this section, I am referring to the lateral and prefrontal cortex and frontopolar cortex. Focal damage to other prefrontal areas, such as the orbitofrontal and medial prefrontal cortex, seem to cause different kinds of memory deficits. I do not feel that the functions of these regions in memory have been adequately sorted out, which is why I do not discuss the literature on these regions. There is an association between medial prefrontal and orbitofrontal damage (typically due to ruptured aneurysms of the anterior communicating artery) and confabulation, which I touch on in a later chapter.

22   These patients formed memories: This pattern of results has been borne out in a number of research studies in patients with focal frontal lesions (Gershberg and Shimamura 1995; Alexander, Stuss, and Gillingham 2009; Stuss et al. 1996; Hirst and Volpe 1988; Della Rocchetta and Milner 1993. See Blumenfeld and Ranganath 2019 for a review.)

24   Activity in the prefrontal cortex was not particularly sensitive: In this sentence, I'm primarily referring to the dorsolateral prefrontal cortex (Brodmann's areas [BA] 9 and 46) and frontopolar cortex (BA 10), which I believe to be the most important for domain general executive functions (cf. Wagner 1999; D'Esposito and Postle 2015). I use the word *particularly* here because researchers have found that particular frontal areas show increased activation during processing of certain types of materials (e.g., Courtney et al. 1998), but the involvement of frontal areas in working memory processes is not nearly as domain-specific as that of posterior cortical areas (Ranganath and D'Esposito 2005; D'Esposito and Postle 2015). The point I'm making here is that the prefrontal cortex isn't "doing" working memory per se, but rather something that contributes to working memory, such as attention, reasoning, planning, and so forth.

24   But the prefrontal cortex was intensely activated: This section provides a brief reference to an incredible period of research from Mark's lab. Some of the papers I'm describing include Druzgal and D'Esposito 2001, 2003; D'Esposito et al. 2006; D'Esposito, Postle, and Rypma 2000; Ranganath, Johnson, and D'Esposito 2000; Ranganath, DeGutis, and D'Esposito 2004; and Ranganath et al. 2004. See Ranganath and D'Esposito 2005, D'Esposito and Postle 2015, Jonides et al. 2008, and especially Badre 2020 for more information on this topic.

24 Instead, fMRI studies and observations of patients supported: The executive analogy was introduced by Pribram (1973) and elaborated by Baddeley and Wilson (1988) and Norman and Shallice (1986). Joaquin Fuster (1980) advanced a slightly different view of the role of the prefrontal cortex in linking perception and action over time. I see that as related to the executive theory.

25 Even in the absence of physical damage: Diamond 2006, West 1996, Moscovitch and Winocur 1992, Craik 1994, and Craik and Grady 2002 provide excellent reviews of developmental and age-related changes in frontal function and/or memory.

25 ADHD is associated with atypical activity: See Arnsten 2009a.

26 one of the first areas to decline: For more extensive coverage of cognitive aging and practical tips, see Levitin 2020 and Budson and Kensinger 2023.

26 This tendency to recall the inane: I'm referring here to the work of Lynn Hasher and Karen Campbell, particularly their studies of "hyper-binding" (see Zacks and Hasher 2006; Campbell, Hasher, and Thomas 2010; and Amer, Campbell, and Hasher 2016).

26 multitasking is probably the most common culprit: See Covre et al. 2019 and Uncapher and Wagner 2018.

26 Our conversations, activities, and meetings are routinely interrupted: For an easy-to-read and practical, science-based discussion of how to deal with information overload, see Levitin 2014.

26 almost always has a cost: I say "almost" because, with extensive practice, two different tasks can be chunked into a single task. For more on how we manage complex tasks, I definitely recommend Badre 2020.

27 "There is no such thing as multitasking": The world of prefrontal cortex research is full of quotable individuals, and Earl Miller is up there with the best.

27 Hypertension and diabetes, for instance, can cause damage: Here, I describe our work with Christine Nordahl, Bill Jagust, and Charles DeCarli (Nordahl et al. 2005, 2006; Lockhart et al. 2012).

27 For instance, people infected with COVID-19: See Douaud et al. 2022 and Becker et al. 2021.

27 The ways in which we neglect: See Krause et al. 2017; Abernathy, Chandler, and Woodward 2010; and Arnsten 2009b for a review.

27 Fortunately, we can do some things to improve: See Voss et al. 2013 and Fillit et al. 2002.

28 One particularly impressive study: Jia et al. 2023.

29 The problem isn't necessarily with the technology: This section on the effects of photography on memory is based on Henkel 2014, Barasch et al. 2017, and Soares and Storm 2018. The literature suggests that taking pho-

tos does not simply have a good or bad effect on memory. It all depends on how you do it. The critical factors involve how you direct your attention and whether you meaningfully engage with the subject matter.

2. TRAVELERS OF TIME AND SPACE

32 memory can be reduced to simple associations between "stimuli": John Watson (1913) pretty much encapsulates what I am talking about here.

33 episodic memory can be differentiated from *semantic memory*: Tulving 1972.

33 a state of consciousness in which we feel: Tulving 1985.

35 "neural networks"—computer programs that mimic: Neural networks were inspired in part by the ideas of a paper by neurophysiologist Warren McCulloch and mathematician Walter Pitts (1943), who modeled a simple network of neurons using electrical circuits. Another key contribution was made by Donald Hebb (1949), a pioneering scientist (and co-adviser of Brenda Milner) who proposed that memories are stored in networks of densely interconnected neurons, and that this happens because learning causes systematic increases and decreases in the strength of connections between specific sets of neurons. Building on Hebb's ideas, Marr (1971) proposed a groundbreaking computational model for "simple memory" based on the biology of the hippocampus. Marr proposed that the hippocampus may be needed to encode specific information and that the neocortex may instead generalize across experiences.

In the early eighties, many psychologists, including Jay McClelland, Donald Rumelhart, and PDP Research Group (1986), used neural networks to explain many learning phenomena. However, in 1988, Gail Carpenter and Stephen Grossberg pointed out a significant challenge, which they called the stability-plasticity dilemma. The dilemma is basically a trade-off between learning new information and losing what was previously learned—how does one design a network that can learn from a single unusual experience without losing all the gains from its previous training? Mike McCloskey and Neal Cohen (1989) ran a systematic set of simulations illustrating the severity of the problem and coined the term *catastrophic interference* (now more widely known as *catastrophic forgetting*). In 1995, Jay McClelland, Bruce McNaughton, and Randy O'Reilly published a paper suggesting that the brain evolved different, "complementary learning systems" to solve the stability-plasticity dilemma. Specifically, they built on Marr's (1971) proposal that the hippocampus can learn quickly from single instances, but it isn't good at generalizing, whereas the neocortex learns slowly but, like traditional neural networks, is good at learning generalities. As I'll talk about later in this book, the authors proposed that the hippocampus can "talk" to the neocortex during sleep, helping the neocortex to learn more quickly without suffering from catastrophic forgetting. In case you are wondering, I have no idea why so many people I have cited here have last names that start with *Mc*.

36 learn the exception to the rule: fMRI research and computational modeling by my grad school classmate Brad Love has shown that the hippocampus is indeed called into action to learn an exception to the rule. See, for example, Love and Medin 1998 and Davis, Love, and Preston 2012.

36 In 1957, she published the paper: A graduate student at McGill University in Montreal at the time, Milner was co-advised by Donald Hebb, who would become a legend in neuroscience, and the brilliant neurosurgeon Wilder Penfield (see Xia 2006 for more information on Milner's work during this period). Penfield used a surgical treatment for epilepsy, cutting off part of the temporal lobe in one hemisphere, thereby removing the area that was causing seizures. Milner first observed memory deficits in patients with temporal lobectomies in her work with Penfield (Penfield and Milner 1958). After they reported their observations at a meeting, neurosurgeon William Scoville contacted Penfield and described his similar experiences. To treat a range of psychiatric and neurological disorders, Scoville had developed a radical procedure, cutting off the temporal lobes in both hemispheres of the brain. He also tried this out on Henry Molaison to treat his severe epilepsy. Scoville invited Milner to study his patients, including H.M., who had severe amnesia (Scoville and Milner 1957). We now know that the one-side temporal lobectomies done by Penfield can actually improve memory if you happen to hit the side of the brain that is producing the seizures. The patients who showed memory problems had tissue removed that was intact. Scoville's procedure always worsened patients' memories because it systematically wiped out both the seizure zone and the tissue on the other side of the brain that the patient had been relying upon. It's akin to removing a flat tire and an intact tire on the other side of the car.

   For more information on H.M.'s remarkable life and impact on the science of memory, I recommend *Permanent Present Tense: The Unforgettable Life of the Amnesic Patient, H.M.*, the 2013 memoir by the late Suzanne Corkin, who worked with Mr. Molaison over her career.

37 five centimeters of tissue: Most neuroscientists erroneously attribute H.M.'s dense amnesia to hippocampal damage, but actually H.M. only lost the anterior two-thirds of his hippocampus, whereas the posterior one-third was spared. He sustained massive damage to the neocortical

gray and white matter (Corkin et al. 1997; Annese et al. 2014), which almost certainly contributed to the severity of his memory deficits.

37 "Memory was the sleeping beauty of the brain": Heidi Roth and Barbara W. Sommer, "Interview with Brenda Milner, Ph.D., Sc.D.," American Academy of Neurology Oral History Project, December 2, 2011, https://www.aan.com/siteassets/home-page/footer/about-the-aan/history/11aantranscriptbrendamilner_ft.pdf.

37 hippocampus must be an all-purpose memory device: This view was articulated most forcefully by Larry Squire, who has argued that the hippocampus is necessary for "declarative memory," which encompasses both new semantic learning and episodic memory (see Squire and Zola 1998 for a review). I agree mostly with the idea that, if you have a hippocampus, you can leverage episodic memory to learn new facts faster. This is basically the argument of McClelland, McNaughton, and O'Reilly 1995, described earlier. However, as I explain later, the perirhinal cortex can support learning of new semantic knowledge.

38 In 1997, Dr. Faraneh Vargha-Khadem: Vargha-Khadem wasn't the first person to study amnesic patients with damage limited (more or less) to the hippocampus, but her report (Vargha-Khadem et al. 1997) was unique in focusing on developmental amnesia. The story about Vargha-Khadem, Jon, and Endel Tulving is described in Vargha-Khadem and Cacucci 2021. Squire and Zola 1998 argued that Jon and other developmental amnesics could learn new semantic knowledge because they have some spared episodic memory capacities, but it does not make sense that these individuals could have such profound amnesia and still learn facts at a rate much faster than adults with hippocampal damage. These individuals are clearly leveraging neocortical plasticity. Squire and Zola 1998 also argued that episodic memory might differ from semantic memory in dependence on the prefrontal cortex, but the preponderance of evidence since the publication of that paper has led to the widespread consensus that the prefrontal cortex is critical for controlled retrieval of both semantic and episodic memories. There is some reason to think that parts of the medial prefrontal cortex contribute to the subjective, first-person experience of mental time travel, but there is unequivocal evidence that the context-specific element of episodic memory depends on the hippocampus, not on the prefrontal cortex (see Ranganath, in press, for a review).

39 We used to think of these moment-to-moment changes as "noise": Jim Haxby and his colleagues at NIMH ran one of the first studies examining the idea that there is useful information in the *pattern* of activity in fMRI data (Haxby et al. 2001). Sean Polyn, Ken Norman, and colleagues

at Princeton applied this insight into memory research in a groundbreaking study in which they used machine learning to mine information from voxel patterns (aka multivoxel pattern analysis or MVPA) to decode information about the context that people were using to recall information from memory (Polyn et al. 2005). Ken Norman saw some data from my student Luke Jenkins and suggested that Luke and I try out a different approach, called representational similarity analysis (RSA; Kriegeskorte, Mur, and Bandettini 2008), which is the "memory code" method that I describe here. RSA is, in my opinion, more interesting than machine learning–based decoding techniques because it gives you a richer amount of information about whether memories that shared similar people, things, or contexts are associated with similar patterns of brain activity. In 2010, Gui Xue and Russ Poldrack at Stanford and our lab concurrently published the first two studies to use RSA to examine episodic memory (Xue et al. 2010; Jenkins and Ranganath 2010).

39 If so, we could use MRI scanning: See Dimsdale-Zucker and Ranganath 2018 for a review of how this works.

39 In our studies, this is exactly what we saw: In this section, I am not describing any single study. The example I give is meant to present, in a simplified manner, the results from a large number of studies that we ran from 2010 to 2020 (Jenkins and Ranganath 2010; Hannula et al. 2013; Hsieh et al. 2014; Ritchey et al. 2015; Libby et al. 2014, 2019; Wang et al. 2016; Dimsdale-Zucker et al. 2018, 2022). Our development of RSA techniques for memory research was spearheaded by an all-star team including Halle Dimsdale-Zucker, Luke Jenkins, Laura Libby, and Frank Hsieh (then students in my lab), and Maureen Ritchey (a postdoctoral researcher at the time, now an accomplished faculty member at Boston College).

40 The hippocampus enables us to "index" memories: I am referring here to hippocampal indexing theory (Teyler and DiScenna 1986; Teyler and Rudy 2007) and cognitive mapping theory (summarized by O'Keefe and Nadel 1979).

40 Because the hippocampus organizes memories according to the context: This finding is called the temporal contiguity effect (Healey, Long, and Kahana 2019), and a number of studies have used this effect to show how the hippocampus organizes episodic memories according to temporal and spatial context. For instance, Miller et al. 2013 actually recorded activity from cells in the hippocampus in epilepsy patients while they navigated in a virtual reality environment. Later on, when the patients recalled events from their VR experience, they activated those cells that fired in the location where those events occurred. See also Umbach et al. 2020 and Yoo

et al. 2021. For convergent evidence from fMRI studies, see Deuker et al. 2016 and Nielson et al. 2015.

41 According to one prominent theory: John O'Keefe of University College London and Lynn Nadel of the University of Arizona introduced this theory, which proposed that the hippocampus evolved to give us a sense of where we are, and that in turn laid the foundation for episodic memory, which is grounded in both space and time (O'Keefe and Nadel 1978). O'Keefe went on to win the Nobel Prize for his discovery of "place cells" in the hippocampus, which activate when an animal is in a particular location in an environment. For a more detailed evolutionary perspective, I strongly recommend reading Murray, Wise, and Graham 2017.

41 I helped Peter develop new memory tests: Peter is a genius. In addition to our study (Cook et al. 2015), he has studied cognition in other species, including even an MRI study of jealousy in dogs (Cook et al. 2018).

42 It's not as if our memories have a time stamp: The relationship between time, space, and episodic memory is reviewed by Ranganath and Hsieh 2016, Eichenbaum 2017, and Ekstrom and Ranganath 2018.

43 Moreover, the environment around us is constantly changing: This is a simplified description of context-based theories of memory (Estes 1955; see Manning, Norman, and Kahana 2014 for a review).

43 My UC Davis colleague Petr Janata: See Janata 2009.

43 recollections of past events in those with Alzheimer's disease: See Baird et al. 2018.

44 Our emotions also contribute to context: The effects of mood on memory seem to be biggest when you are trying to recall an event without any cues, and when your emotion is salient and central to the event (Bower 1981; Eich 1995).

45 Just as being in a particular place: See Mandler 1980.

45 adults do not have reliable episodic memories: Sometimes people do "remember" events from their first few years, not because they are mentally traveling back to that time but rather because they generated a memory from seeing pictures and hearing stories from family members. See Peterson 2002, Howe and Courage 1993, and Bauer 2004 for more on the topic of infantile amnesia.

45 based on research by my UC Davis colleague Simona Ghetti: See Ghetti 2017.

45 connections between neurons across the entire neocortex: Johnson 2001.

46 We naturally update our sense of context: See Zacks and Tversky 2001.

46 The context change that occurs with an event boundary: See Radvansky and Zacks 2017 for a review.

46 Recent work from a number of labs: This sentence summarizes a lot of

work. Swallow et al. 2009, 2011 found that objects shown at event boundaries (i.e., the end of an event) in a film were better remembered than objects shown in the middle of an event. Aya Ben-Yakov (Ben-Yakov, Eshel, and Dudai 2013) found that hippocampal activity at the *end* of a short film was predictive of successful encoding of the film. Baldassano et al. 2017 built on this finding by using machine learning to show that the DMN (default mode network) exhibits dramatic shifts in activity patterns (i.e., memory codes) around event boundaries and that hippocampal activity spikes when the DMN activity pattern shifts (see also Ben-Yakov and Henson 2018). Zach Reagh in my lab found that hippocampal activity spiked at human-identified event boundaries (rather than moments of DMN activity pattern shifts) and that the hippocampal boundary response predicted individual differences in memory performance on an entirely different test (Reagh et al. 2020). Alex Barnett in my lab (Barnett et al. 2022) found that hippocampal activity and functional connectivity with the posterior medial subnetwork of the DMN predicted successful encoding of the preceding event. Finally, Lu, Hasson, and Norman 2022 demonstrated with a simple neural network model that it may be optimal for the brain to encode episodic memories at the end, rather than in the middle, of an event.

47   It can be frustrating—even alarming: Jeff Zacks (2020) provides an excellent review of the effect of event boundaries on memory, including aging-related effects. Analyzing a huge amount of fMRI data collected in the UK, Zach Reagh and I (Reagh et al. 2020) found that the degree to which the hippocampus lights up at an event boundary predicts individual differences in episodic memory, and overall activation at event boundaries decreases with age.

47   On average, people find it easier to recall positive experiences: See Adler and Pansky 2020 for more on positive memory biases and Mather and Carstensen 2005 for more on aging and positivity effects in memory.

48   memories from these years is called the *reminiscence bump*: On the reminiscence bump, see Jansari and Parkin 1996; Krumhansl and Zupnick 2013; Schulkind, Hennis, and Rubin 1999; and Janssen, Jaap, and Meeter 2008.

48   The term *nostalgia* was coined by a Swiss physician: Anspach 1934.

48   if people felt lonely in their daily lives: Newman and Sachs 2020.

48   having Highly Superior Autobiographical Memory: A special group of people with Highly Superior Autobiographical Memory seem to have a precisely dated mental catalog of minutiae of their lives (see Leport et al. 2012, 2016). Oddly, they do not seem to be any different from anyone else at memorization or other lab studies of memory. Others, with Severely Deficient Autobiographical Memory (SDAM), seem to have almost no capability to relive their personal past, and these individuals don't tend to

ruminate (Palombo et al. 2015). Brian Levine, one of the world's foremost authorities on autobiographical memory, thinks that I have SDAM, but sadly I do seem to ruminate a lot.

48 "I do tend to dwell on things": MacMillan 2017.

49 Recalling negative events can remind us: See De Brigard and Parikh 2019 for a perspective on how this process, termed *episodic counterfactual thinking*, can work as a productive emotion regulation strategy.

49 In one study, people who vividly recalled: See Gaesser and Schacter 2014.

3. REDUCE, REUSE, RECYCLE

51 Scott turned to the science of memory: Hagwood 2006.

51 a book by British memory trainer Tony Buzan: Buzan 1984.

51 Mongolian Swedish triple world-record holder: *Memory Games*, documentary directed by Janet Tobias and Claus Wehlisch (Stockholm: Momento Film, 2017); Lakshmi Gandhi, "Memory Champion Yänjaa Wintersoul Believes Anyone Can Learn to Remember," NBC News, November 2018.

51 Yänjaa is perhaps most famous for a 2017 viral video: Yänjaa Wintersoul, *The IKEA Human Catalogue Test* (advertising campaign), art directed by Kooichi Chee and Jon Loke (UK: FreeFlow/Facebook Creative, 2018); *The IKEA Human Catalogue Test (Extended)*, YouTube.

52 In 1956, George Miller: Miller 1956. I suspect that the "famous senator" described by Miller is Senator Joe McCarthy, who stoked public paranoia about the threat of communists in the United States.

52 the human brain can only keep a limited amount: There is some debate about the exact number, or the extent to which such a number even exists. George Miller (1956) famously suggested seven, plus or minus two as the "magic number." I cite the more recent estimate of three, plus or minus one, which is based on more precisely controlled studies (see Cowan 2010; Luck and Vogel 2013). Others disagree, saying that you can't put a number on it at all. I am not sure of the answer, but it doesn't really matter. Everyone agrees that there are limits to the amount of information that we can hold in mind.

53 Chunking allows us to compress: Miller (1956) used the term *recoding* to describe what we now call *chunking*. The latter term was used by Herb Simon (1974).

53 Simon first became interested: Newell, Shaw, and Simon 1958.

54 When Simon studied chess experts: Chase and Simon 1973.

54 Like memory athletes, chess grand masters: Simon originally espoused the idea that expertise changes how grand masters perceive the board and chess pieces (Chase and Simon 1973). Later, as his research evolved,

he came to believe that expertise affects how people store information in memory, so that they can build more sophisticated templates that exploit the structure of the game (Gobet and Simon 1998). See also Ericsson and Kintsch 1995.

54 At the time, most neuroscientists believed: See Bukach, Gauthier, and Tarr 2006 for a review.

55 MRI scans showed that activity in the prefrontal cortex: Moore, Cohen, and Ranganath 2006.

56 Instead of focusing on his dissertation research: You can find some of his archived posts at http://baseballanalysts.com/.

56 His work on predictive analytics: Chris was billed by his alma mater at a speaking event in December 2016, following the Cubs' World Series win, as "The Princeton Club of Chicago invites fellow Tigers and Cubs fans to learn about the 'math' behind the 'magic' from 'Moneyball Man' Chris Moore."

56 we can exploit one of the brain's most powerful tools: Bartlett (1932) introduced the term *schema* as it is used in memory research. Kant (1899) introduced *schema* in his *Critique of Pure Reason*. Piaget (1952) proposed how schemas operate in cognitive development, and David Rumelhart (e.g., Rumelhart and Ortony 1977) elaborated on schemas in artificial intelligence and memory. Other relevant work includes Minsky's Frame Theory (1975), and Schank and Abelson's Script Theory (1977). As an aside, I use *schemas* rather than the more pretentious plural form *schemata*.

57 mental map for the mazelike layout: I refer here to Tolman's (1948) concept of the *cognitive map*. Tolman introduced this term as a way of capturing how animals can navigate through space with a compact mental representation of the relationships between the important points in an environment. Tolman's description is, in my opinion, very much in the spirit of Bartlett's schemas. Today, scientists and science journalists mistakenly interpret Tolman's cognitive map as a literal, euclidean map of space (or a two-dimensional coordinate space to represent anything). That entirely misses the point of Tolman's paper.

57 We all have mental blueprints: Franklin et al. 2020 review some of the work in this area, with a computational model for how event schemas are learned and used.

59 the neocortex is organized into networks: Malcolm Young (the neuroscientist, not the late AC/DC guitarist) deserves credit for introducing computational techniques to identify networks in the neocortex (e.g., Hilgetag O'Neill, and Young 1997). Olaf Sporns (2010) provides a fantastic, readable overview on brain networks.

59 Raichle proposed that this network of areas: See Raichle et al. 2001.

60 The DMN is often studied: See, for example, Mason et al. 2007 and Small-
    wood and Schooler 2015. I don't mean to imply in any way that it is wrong
    to talk about these topics in relation to the DMN. My point here is that,
    in addition to playing a role in imagination, mind-wandering, and self-
    reflection, the DMN contributes positively to many kinds of high-level
    cognitive tasks that resemble what we do in the real world, such as auto-
    biographical memory retrieval, spatial navigation, and reasoning. But if
    you want to know more about this topic, I suggest reading cognitive neu-
    roscientist Moshe Bar's *Mindwandering* (2022), a terrific and easy-to-read
    book. Although he does talk about the role of the DMN in daydreaming,
    Bar's research heavily influenced my thinking about the functions of the
    DMN.

60 I became clued in to a growing number of fMRI studies: I credit a con-
    versation with Mick Rugg, a pioneering cognitive neuroscientist, as the
    inspiration for sending me down the DMN rabbit hole. Mick had just done
    an influential review paper demonstrating that the entire DMN shows
    increased activation when we recollect words from a study list (Rugg and
    Vilberg 2013), and he pointed out the parallel between what everyone was
    seeing (and overlooking) in fMRI studies of memory and what everyone
    was seeing in fMRI studies of brain networks. Mick pointed me to great
    work by Randy Buckner, Jess Andrews-Hanna, and Daniel Schacter (2008)
    that captured these parallels.

60 We proposed that cell assemblies: Ranganath and Ritchey 2012.

61 I started to see new results: When I say "new" results, I mean that the
    results from the Doeller (Milivojevic et al. 2015, 2016) and Norman/
    Hasson labs (Chen et al. 2017; Baldassano et al. 2017, 2018) got my atten-
    tion, but other innovative fMRI studies of memory for naturalistic stimuli
    had set the precedent for thinking about these issues (e.g., Zacks et al.
    2001; Swallow, Zacks, and Abrams 2009; Ezzyat and Davachi 2011).

61 I was sufficiently inspired: In our first joint publication, Nick Franklin (then
    a postdoc in Sam Gershman's lab) introduced an ambitious computational
    model to explain schemas, event boundaries, and reconstruction of epi-
    sodic memories (Franklin et al. 2020).

61 We transitioned from studying brain activity: See Barnett et al. 2022;
    Reagh et al. 2020; Cohn-Sheehy et al. 2021, 2022; and Reagh and Ranga-
    nath 2023.

62 When the experiment was done: These results are from Reagh and Ran-
    ganath (2023). I've simplified the description, but I encourage the reader
    to read this study for more information. We actually found differences
    across three different subnetworks of the DMN. The Posterior Medial
    Network (PMN) encoded different memory codes for each café and each

supermarket; the Medial Prefrontal Network (MPN) encoded a generic memory code for supermarket movies and a generic memory code for café movies; and the Anterior Temporal Network (ATN) encoded a separate memory code for each character.

63 hippocampal activation spikes as it communicates: Reagh et al. 2020 showed that hippocampal activity spikes at event boundaries during a film, and Barnett et al. 2022 showed that hippocampal functional connectivity with the DMN at event boundaries predicted successful encoding of memories for the event. Note that overall DMN-hippocampal connectivity did not spike at event boundaries, but the connectivity increase was apparent for successfully encoded events, relative to events that were not recalled on the subsequent memory test.

63 amyloid—a protein implicated in the development of Alzheimer's disease: Palmqvist et al. 2017.

65 "when you have a high basketball IQ": Melissa Rohlin, "Inside the Mind of LeBron James: An Exclusive Look at His Basketball IQ," *Sports Illustrated*, March 27, 2020, https://www.si.com/nba/2020/03/27/inside-the-mind-of-lebron-james-a-look-at-his-iq.

65 "You better save your favorite play": Brian Windhorst, "Total Recall: LeBron's Mighty Mind," ABC News, July 22, 2014, https://abcnews.go.com/Sports/total-recall-lebrons-mighty-mind/story?id=24662461.

4. JUST MY IMAGINATION

70 In his classic 1968 monograph: Luria 1968, 11.

70 *New Yorker* writer Reed Johnson: Johnson 2017.

71 enabled me to elaborate: This is a simplified presentation, and interested readers can check out Sheldon, Fenerci, and Gurguryan 2019 and St. Jacques 2012 for more detailed perspectives on the topic.

71 the brain activity changes that occur: Addis, Wong, and Schacter 2007; Szpunar, Watson, and McDermot 2007; Hassabis, Kumaran, and Maguire 2007.

72 *Science* magazine declared it: Miller 2007.

72 Bartlett began his research on human memory: My biographical description of Bartlett is drawn from Roediger 2003.

72 he focused not on memory: Bartlett 1928a, 2014.

72 his most important work: Bartlett 1932.

73 in 2015, Williams recounted: Amanda Taub, "The Brian Williams Helicopter Scandal: A Clear Timeline," *Vox*, February 9, 2015, https://www.vox.com/2015/2/5/7987439/brian-williams-iraq-apology-helicopter.

74 an experiment that is now taught: The study (Roediger and McDermott

1995) is usually taught as the "DRM" effect because their procedure was adapted from Deese 1959.

75 Individuals with amnesia generally don't fall: Schacter, Verfaellie, and Pradere 1996.

75 people with autism or other neurodevelopmental disorders: Griego et al. 2019, Beversdorf et al. 2000. But see Solomon et al. 2019 for findings that those with ASD can sometimes show a typical pattern of false memory effects.

76 We like to think of memories as though they are tangible: Neuroscientists are no different. Many scientists have sought to identify the neural substrates of individual memories, also known as engrams (e.g., Josselyn, Köhler, and Frankland 2017). I feel that the search for engrams in neuroscience reflects an underlying assumption that memories are static, objective records of the past, whereas I share Bartlett's belief that the same experience can be used to construct a nearly infinite range of memories.

77 assumptions about people's motivations: Owens, Bower, and Black 1979.

77 perspectives will also shape how they reconstruct that event: Pichert and Anderson 1977.

77 rival fans of two German soccer teams: Huff et al. 2017.

77 Shifting perspective can also help: Anderson and Pichert 1978.

77 she had a group of volunteers: Loftus and Palmer 1974.

78 we wonder what our present would be like: My friend Felipe De Brigard of Duke University is an expert on this topic. He has shown, among other things, that the DMN is engaged when we do this kind of what-if thinking. De Brigard and Parikh 2019 provides a nice synopsis of the emerging research in this area.

79 consciously applying some critical thinking: Johnson and Raye 1981. To keep things intuitive and simple, I did not mention that Marcia expanded her framework to encompass how we think about the sources of any conscious experience, otherwise known as source monitoring (Johnson, Hashtroudi, and Lindsay 1993).

79 imagined events are more focused: Johnson et al. 1988. On a related note, when I was in graduate school, my officemate (and die-hard Boston Red Sox fan) Brian Gonsalves and our adviser, Ken Paller, found neural evidence for this idea as well (Gonsalves and Paller 2000). Remembering an object that you recently saw will trigger activity in visual areas of the brain—much more so than if you simply imagined that object (see also Wheeler, Petersen, and Buckner 2000; Nyberg et al. 2000).

80 Anything that affects the prefrontal cortex: Simons, Garrison, and Johnson 2017 provides a nice review of the topic.

80 Marcia Johnson's ideas inspired my dissertation research: In my dissertation research (Ranganath and Paller 1999, 2000), I used event-related potentials (ERPs), which gave us a clear picture of when and how people monitored the accuracy of their memories, but they did not permit us to figure out exactly which parts of the prefrontal cortex were activating when people engaged in source monitoring. My fMRI studies with Marcia and with Mark D'Esposito (Ranganath, Johnson, and D'Esposito 2000, 2003) showed these areas.

80 Our findings coincided with similar results: See, for example, Dobbins et al. 2002; Schacter et al. 1997; Wilding and Rugg 1996; Cabeza et al. 2001; Johnson, Kounios, and Nolde 1997; Johnson et al. 1997; Curran et al. 2001. Rugg and Wilding 2000 provides an excellent review of the emerging cognitive neuroscience literature during this time.

80 people who have more gray matter: Buda et al. 2011.

80 some people who have extensive damage: In my past work as a clinician, I found that confabulation is rare and usually transient. The cases that have been reported in the literature usually have some combination of amnesia, due to diencephalic damage, and deficits in executive functions, due to damage in the medial prefrontal and orbitofrontal cortex. My thinking on this topic was heavily influenced by a chapter on confabulation by Morris Moscovitch (1989). Johnson et al. 2000 also reviews confabulation.

80 Neurologist and author Jules Montague: Montague 2019.

80 As we get older, prefrontal function gets worse: Hashtroudi, Johnson, and Chrosniak 1989.

81 If we visualize something in extraordinary detail: Gonsalves et al. 2004; Thomas, Bulevich, and Loftus 2003.

81 Now, consider how one individual: Hassabis et al. 2007. I should add that Maguire, Vargha-Khadem, and Hassabis 2010 found other people with amnesia who have no problem with imagination. I suspect that those who have an inability to imagine the future have a disconnection between the hippocampus and the DMN (Barnett et al. 2021).

82 In a 1928 paper that was far ahead: Bartlett 1928b.

82 the hippocampus and the DMN might function at the crossroads between memory and imagination: Schacter and Addis 2007; Schacter, Addis, and Buckner 2008.

82 laboratory tasks designed to elicit creative thinking: Beaty et al. 2018, Madore et al. 2019.

82 dysfunction in these areas: Duff et al. 2013, Thakral et al. 2020.

82 people who show higher performance: Thakral et al. 2021.

83 "We can only create a collage": Austin Kleon, "Re-imagining from Mem-

ory," March 26, 2008, https://austinkleon.com/2008/03/26/re-imagining
-from-memory/.

84 Unlike a computer program: I recognize that a number of advanced AI
applications are trained on massive datasets that are, in a sense, diverse.
However, the internet content that is available for training generative AI
models is not a random sample of human works, and certain demographic
groups are highly overrepresented. Moreover, these programs are often
trained for a particular task, such as predicting the next word in a sentence
or assigning labels to images. That is a bit different from a human who
is exposed to stimuli without any particular goal. Human artists can also
draw creative parallels between eclectic sources of inspiration that might
be treated as fundamentally different by an AI product. Having said all
that, I do think that human artists can use generative AI to create innovative art, much as visual artists have used collage and hip-hop artists use
dense layers of samples from existing works.

84 Pablo Picasso's cubist paintings were preceded: Scenic, "Picasso's Greatest
Influences Explored: From the Blue Period to Cubism," November 3, 2020,
https://www.scenic.co.uk/news/picassos-greatest-influences-explored
-from-the-blue-period-to-cubism.

84 Akira Kurosawa, who spent his formative years: Peggy Chiao, "Kurosawa's
Early Influences, Criterion Collection, October 19, 2010, https://www
.criterion.com/current/posts/444-kurosawa-s-early-influences#:~:text=
Another%20major%20influence%20on%20Kurosawa,Vsevolod%20
Pudovkin%2C%20and%20Sergei%20Eisenstein.

84 Members of the rap group Wu-Tang Clan: Ghansah 2011.

5. MORE THAN A FEELING

87 Emotions, the conscious feelings we experience: This is a complicated
topic that is intensely debated among social, personality, and clinical psychologists, affective neuroscientists, and even anthropologists. This is my
own take, which is influenced by NYU neuroscientist (and musician) Joe
LeDoux. LeDoux (2012) differentiates between "survival circuits," which
I'm essentially describing as motivational circuits, and conscious feelings.

88 Our emotions . . . are shaped: Often but not always. Again, the relationship between emotions and survival circuits is complicated, and there are
several points of view (see Adolphs, Mlodinow, and Barrett 2019; Gross
and Feldman Barrett 2011).

88 When a survival circuit: See Avery and Krichmar 2017 for a review.

88 Some neuromodulators are like the hot sauce: See Nadim and Bucher

2014 for a review. As an aside, I should clarify that, in this chapter, I will be discussing evidence about the roles of different neuromodulators in particular memory processes. My goal was to convey the key findings in a simple, readable way, but I hope that advanced readers will appreciate that neuromodulators are complex, and different neuromodulatory systems interact with one another (for instance, neurons in the locus coeruleus release both norepinephrine and dopamine). Accordingly, it is highly unlikely that any neuromodulator works in isolation to perform a single function in the human brain.

88 Neuromodulators also promote *plasticity*: See Kandel, Dudai, and Mayford 2014, McGaugh 2018, and Takeuchi, Duszkiewicz, and Morris 2014 for a review.

89 emotional arousal ratchets up the stakes: See Mather 2007 and Mather et al. 2016. One additional note: everything I say about noradrenaline applies both to threatening experiences and to exhilarating experiences that get you aroused.

89 it should change *what* we will remember: Mather 2007.

89 The effects of noradrenaline keep going: See McGaugh 2018 for a review. One terminology note: When memory for an event is affected by a drug or neuromodulator after the event is over, neuroscientists say that the chemical is affecting "memory consolidation." I've avoided getting into that terminology here, mostly because the term has been used in the research literature to refer to many different things, all of which carry unnecessary baggage. In practice, neuromodulators affect memory encoding as well as consolidation.

90 When we recall a time: See Phelps 2004 and LaBar and Cabeza 2006. For an example of one study making this point, see Ritchey et al. 2019. Also, I have emphasized the hippocampus, but Yonelinas and Ritchey 2015 makes an excellent case for teamwork between the amygdala and perirhinal cortex, though I do not have space to get into it here.

90 people with amygdala damage: Adolphs et al. 1997.

90 Conversely, people with damage to the hippocampus: See Bechara et al. 1995.

91 Individuals who have endured extreme trauma: See Lensvelt-Mulders et al. 2008.

91 This is especially true: See McEwen 2007, Sapolsky 2002, and Godoy et al. 2018.

91 think something bad might happen: Credit to Sue Mineka's graduate seminar in psychopathology research (see Mineka and Kihlstrom 1978 for a review).

92 hippocampus, prefrontal cortex, and amygdala all have: McEwen, Weiss, and Schwartz 1968.

92 the effects of stress on memory: Shields et al. 2017; Wolf 2009; Sazma, Shields, and Yonelinas 2019.

92 To study the ways: In the next few paragraphs, I'm describing a range of findings on stress summarized by Shields et al. 2017. The reference for Andy's crazy skydiver experiment is Yonelinas et al. 2011; as a bonus, it is published in a journal called *Stress*.

92 stress hormones seem to promote plasticity: Reviewed by McIntyre, McGaugh, and Williams 2012.

92 a mechanism that enhances memory: See Sazma, Shields, and Yonelinas 2019 for a review.

92 effects of acute stress in rats or mice: All the scientific studies and experiments on animals referenced in this book were subject to oversight by the IACUC (Institutional Animal Care and Use Committee), which ensures that the highest standards of animal welfare are maintained.

92 more isn't always better: Sapolsky 2003 provides a nice review of this issue. We (McCullough et al. 2015; Ritchey et al. 2017; Shields et al. 2019) have also studied how individual differences in stress responses (measured via cortisol levels) affect memory.

93 Stress tips the chemical balance: Amy Arnsten (2009b), an expert on neuromodulation in the prefrontal cortex, provides an outstanding review of the topic here based on work in animal models. See Shields, Sazma, and Yonelinas 2016 for a review of the literature in humans.

93 anyone who spends sustained time: Mineka and Kihlstrom 1978 discusses the relevance of predictability and controllability on anxiety. Lupien et al. 2009 reviews the neurotoxic effects of chronic stress.

93 Over time, the cumulative impact of stress: Liberzon and Abelson 2016 provides a wonderful review of the literature on PTSD and memory. My discussion in this section is based in part on the authors' theory that PTSD is caused by a loss of context specificity of hippocampally dependent memories.

94 disorder known as dissociative fugue: Markowitsch 2003 covers the neuroscience of fugue states, along with other forms of so-called psychogenic amnesias.

94 A fugue episode is thought to explain: Stefania de Vito and Sergio Della Sala, "Was Agatha Christie's Mysterious Amnesia Real or Revenge on Her Cheating Spouse?," *Scientific American*, August 2, 2017, https://www.scientificamerican.com/article/was-agatha-christie-rsquo-s-mysterious-amnesia-real-or-revenge-on-her-cheating-spouse/.

94 the majority of patients appear: Harrison et al. 2017, Staniloiu and Markowitsch 2014.

95 The most reliable way to (ethically) stress: Kirschbaum, Pirke, and Hellhammer 1993.

95 being in an unstable social hierarchy: Knight and Mehta 2017.

95 Spending a long time in such an environment: I strongly recommend that readers who want more on this topic check out Robert Sapolsky's wonderful 1994 book, *Why Zebras Don't Get Ulcers*. For a more academic reference, see Dickerson and Kemeny 2004.

96 Dopamine is about *motivating*: Wise 2004; Robbins and Everitt 2007.

96 Berridge demonstrated that: This work was reviewed in Berridge and Robinson 2016.

96 Dopamine in the amygdala: Schultz 1997.

96 in a brain region called the *nucleus accumbens*: Haber 2011.

96 Because dopamine helps us form memories: Pennartz et al. 2011.

96 we are wired to learn only: Schultz 2006.

98 In our gambling studies: Cohen and Ranganath 2005, 2007; Cohen, Elger, and Ranganath 2007.

99 If a person had a large neural response: Cohen and Ranganath 2007 and Cohen, Elger, and Ranganath 2007 used electroencephalography to measure neural responses to rewarding outcomes.

99 One of the unexpected discoveries: See, for example, Cohen et al. 2005, Cohen and Ranganath 2005, Cohen 2007, Knutson et al. 2001.

100 consciously remember the feelings that compelled us: I am referring here to the "empathy gap." I was influenced by Shankar Vedantam's *Hidden Brain* podcast, December 2, 2019, https://www.npr.org/2019 /11/27/783495595/in-the-heat-of-the-moment-how-intense-emotions -transform-us. See Loewenstein and Schkade 1999 for a review of the relevant academic literature.

100 drugs that hijack our survival circuitry: See, for instance, Wise and Robble 2020, Volkow et al. 2007. However, the causes of addiction are multifactorial, and it would be remiss to ignore the contributions of psychosocial factors (e.g., Hart 2017; Heilig et al. 2021).

100 For some people, these effects are transient: Here, I refer to findings indicating that not everyone who uses "hard drugs" develops a full-blown addiction (Schlag 2020). Ahmed 2010 reviews the relevant literature based on studies in rats and epidemiological studies in humans.

101 matter of memory triggered by context: Perry et al. 2014.

101 Pretty much every living thing: My friend Sam Gershman and others (2021) wrote an interesting review on this topic.

6. ALL AROUND ME ARE FAMILIAR FACES

104 translates into English as "already seen,": Anne Cleary is one of the world's experts on this topic. See Cleary and Brown 2021 for a deep, scholarly, and fascinating review on this topic, drawing on work in history, religion, philosophy, psychology, and neuroscience.

104 Déjà vu is an almost universal: I have read that two out of three people report having had at least one déjà vu experience; however, I have been unable to find a solid source for that estimate. My friend and colleague Nigel Pedersen, who studies the phenomenon (and had many helpful suggestions on this section), says that this number probably underestimates the frequency of déjà vu because people often ignore these experiences when they happen or forget them soon after. In an informal poll of students in my human memory class, almost everyone indicated having experienced déjà vu or something related (there are many French words for various "déjà" phenomena). So, I'm going with "almost universal," but your experience may vary.

105 Penfield used an electrical brain-stimulation: Penfield learned this technique while working with Otfrid Foerster and reported the findings from several cases in later papers (Foerster and Penfield 1930; Penfield 1958).

105 when Penfield stimulated areas: Penfield 1958; Mullan and Penfield 1959. I should also mention that some individuals with temporal lobe epilepsy have a strong déjà vu sensation as part of the "aura" that precedes a seizure as electrical impulses begin firing off in this part of the brain (Hughlings-Jackson 1888).

105 Rebecca Burwell: See Ho et al. 2015.

107 our sense of familiarity reflects: Ebbinghaus (1885) noted that some experiences can lead to learning even if they cannot be recollected: "These experiences remain concealed from consciousness and yet produce an effect which is significant and which authenticates their previous experience." Larry Squire (1986), Endel Tulving and Daniel Schacter (Tulving and Schacter 1990), and others have endorsed various versions of the "multiple memory systems" hypothesis, which suggests that there are separate systems for conscious (also known as declarative or explicit) and unconscious (also known as nondeclarative, procedural, or implicit) expressions of memory. Henry Roediger (1990) has argued that exposure to something can help you process it more easily (aka fluency) the next time around. In this view, fluent processing of stimuli can sometimes happen without awareness, and under other conditions, people can become aware of those effects, even if they have no episodic memory for the stimulus. It is certainly true that people can show more fluent process-

ing of words or objects without any awareness that they have seen those items recently. There's also a general consensus that the guts of episodic memories—information about who, what, where, when, and how—are stored in the neocortex, and that learning by neocortical cell assemblies is also what drives priming. As a result, others such as Neal Cohen and Howard Eichenbaum (1995) proposed that it might make more sense to distinguish between the kind of fluency effects that come about from learning that happens in a particular set of neocortical areas and the ability to remember events via interactions between the hippocampus and neocortex. My take is a hybrid of Roediger's and Cohen and Eichenbaum's ideas.

107 neuroscientists such as the late Mort Mishkin: Mortimer Mishkin was a pioneering neuroscientist working at the U.S. National Institute of Mental Health, studying the neural basis of memory in monkeys. He was impressed by the work of Vargha-Khadem on developmental amnesia and, in 1997, boldly proposed that the perirhinal cortex can be sufficient to support acquisition of semantic knowledge, whereas the hippocampus is needed for episodic memory (Mishkin et al. 1997). Aggleton and Brown (1999), who studied memory primarily in monkeys and rats, deserve considerable credit for taking this idea one step further, by suggesting that the perirhinal cortex is also sufficient to support memory based on familiarity. I also strongly recommend reading O'Reilly and Norman 2002 for an accessible review of how computational models can give insights into the different contributions of the perirhinal cortex and the hippocampus to familiarity and episodic memory.

107 familiarity can be strong: Andy Yonelinas suggested that proper episodic memory (aka recollection) could be discriminated from familiarity based on people's subjective experience (Yonelinas 2001). One of Andy's methods was to ask people to recognize words or pictures that they had seen before, then to have the subjects rate how certain they were for each memory judgment (i.e., confidence) on a 1–6 scale (Yonelinas 1994). When they had an episodic memory experience and recollected the item, they would respond with a 6, meaning that they were absolutely certain it was an "old" word. For instance, if they saw *blueberry* on the test and remembered craving blueberry pancakes while studying the word, that would lead them to give it a 6. But when they did not remember anything specific about a word they had studied, they might rate their confidence 4 or 5—in effect, indicating they were guessing or following a hunch—but more often than not, those guesses were correct. Andy believed people were making those "guesses" based on a sense of familiarity. Andy and his UC Davis colleague Neal Kroll ran a groundbreaking study showing that patients with dam-

age to the hippocampus could still recognize words that they had studied based on familiarity, but they lost the ability to recognize words based on recollection (Yonelinas et al. 2002). The method and model that Andy used became the basis for our beer-bet study, as well as a number of other studies that came afterward.

There are numerous debates in the field regarding the details of Andy's approach to modeling of recollection and familiarity from confidence ratings (Yonelinas and Parks 2007). Some of the specific criticisms of Andy's "receiver operating characteristics" modeling procedure are legitimate (e.g., Wixted 2007), but they do not nullify the larger point that recollection and familiarity are different. Unfortunately, psychology tends to sentence ambitious ideas to death by a thousand cuts. For a taste of this debate, see Wixted 2007, Yonelinas 2002, and Yonelinas et al. 2010.

108 we decided to team up on an experiment: While our study (Ranganath et al. 2004) was sitting in the purgatory of peer review, Lila Davachi, Jason Mitchell, and Anthony Wagner (2003) published a study showing that perirhinal cortex activity was sufficient to support recognition of familiar words, and that hippocampal activity was more specifically related to putting together those words with contextual information. Both of our studies, and a host of subsequent fMRI studies, made the same point, showing that the hippocampus seemed to encode memories that specifically told you about the context in which you saw a particular face or word, consistent with Aggleton and Brown 1999. Montaldi et al. 2006 is a particularly rigorous (and underappreciated) study that hammers home this point.

110 familiarity isn't a weak form of episodic memory: Our review (Eichenbaum, Yonelinas, and Ranganath 2007) supported the idea that memory could be differentiated into three different components—information about items (people and things), contexts (places and situations), and associations between items and contexts (episodic memories)—and that the "strength" of item memories is what gives rise to the feeling of familiarity. Our Binding of Items and Contexts (BIC) model proposed that the perirhinal cortex supports item memory representations, the parahippocampal cortex supports context representations, and the hippocampus binds item and context information. This was a fairly straightforward extension of Mishkin et al. 1997's framework, which linked the perirhinal cortex to semantic memory, the parahippocampal cortex to spatial memory, and the hippocampus to spatial and episodic memory. Eacott and Gaffan 2005 and Knierim, Lee, and Hargreaves 2006 summarized the relevant work from animal models, and Davachi 2006 reviewed work from fMRI studies in support of a model that is pretty much the same as what we proposed. The relatively unique contribution of our paper was to bring all this evidence

together and to relate it to people's conscious experiences—familiarity and recollection—rather than speaking strictly about memory for items and contexts alone. Notably, the latter approach is problematic because people can recognize an item based on recollection and perform contextual memory decisions based on familiarity. Thus, it was important to distinguish between the *processes* of recollection and familiarity, rather than the kind of memory *test* that is given by the experimenter. For further reading, see Diana, Yonelinas, and Ranganath 2007; Ranganath 2010; Mayes, Montaldi, and Migo 2007; and Norman 2010.

111   Those little tweaks also happen: See Mishkin et al. 1997; Aggleton and Brown 1999; Ranganath and Rainer 2003; O'Reilly and Norman 2002; Grill-Spector, Henson, and Martin 2006; Eichenbaum, Yonelinas, and Ranganath 2007; and Ranganath 2010 for various versions of this hypothesis.

111   In our fMRI studies: Here, I am discussing work by Wei-Chun Wang, a graduate student working with Andy Yonelinas, who collaborated with me on two fMRI studies examining the relationship between familiarity and "conceptual priming," which is the ability to fluently access the meaning of a word when it is repeatedly processed. In our first study, Wei-Chun showed that activity increases in the perirhinal cortex during encoding of words predicted conceptual priming on a subsequent memory test (Wang et al. 2010). Wei-Chun also found that patients who had left-side temporal lobectomies did not show conceptual priming if the lesion impinged on the perirhinal cortex. In a subsequent fMRI study of activity during the *retrieval phase*, he found that activity *reductions* in the perirhinal cortex predicted a greater degree of conceptual priming and higher levels of familiarity (Wang, Ranganath, and Yonelinas 2014). The latter finding is in accord with neurophysiological recordings from the perirhinal cortex by Malcolm Brown in monkeys (Brown and Xiang 1998) and intracranial recordings from the perirhinal cortex by Kia Nobre and Greg McCarthy (1995) in humans.

I do not mean to imply that all forms of implicit memory are dependent on the perirhinal cortex (they are not) nor that familiarity is always associated with the perirhinal cortex (no one knows for sure). Paller, Voss, and Boehm 2007 presents an alternate perspective that captures some of the gray areas that I do not get into here.

111   when the name of the actor: The classic experiment on the tip-of-the-tongue effect came from Brown and McNeill 1966, and Brown 1991 has a thorough review of the many studies that followed, including the "blocking" hypothesis that I present in this paragraph. Maril, Wagner, and Schacter 2001 found that having something on the tip of the tongue seems to ramp up activity in the prefrontal cortex, which makes sense, because

once you feel that something is on the tip of your tongue, you end up putting all your mental resources to work to find that memory.

112 Familiarity can bubble to the surface: See Kelley and Jacoby 1990.

112 a phenomenon scientists call the "mere exposure effect": Bob Zajonc (1968) first reported the mere exposure effect. See Zajonc 2001 for a review.

112 In *cryptomnesia*, sometimes referred to as: Brown and Murphy 1989 has a nice experimental demonstration of this effect.

112 its success also led to a lawsuit: "George Harrison Guilty of Plagiarizing Subconsciously a '62 Tune for a '70 Hit," *New York Times*, September 8, 1976, https://www.nytimes.com/1976/09/08/archives/george-harrison-guilty-of-plagiarizing-subconsciously-a-62-tune-for.html.

112 "Did Harrison deliberately use the music": Bright Tunes Music Corp., Plaintiff, v. Harrisongs Music, Ltd., et al., Defendants, 181.

112 "because it already had worked": "George Harrison Guilty of Plagiarizing Subconsciously a '62 Tune for a '70 Hit."

113 "99 percent of the popular music": Harrison 2007, 340.

113 Richard Nisbett and Timothy Wilson set out to explore: See Nisbett and Wilson 1977.

114 Consider the case of Robert Julian-Borchak Williams: Kashmir Hill, "Wrongfully Accused by an Algorithm," *New York Times*, June 24, 2020, https://www.nytimes.com/2020/06/24/technology/facial-recognition-arrest.html.

116 1.4 times better at recognizing: Meissner and Brigham 2001. I must stress, however, that race is a social construct. Although race biases have been found in a number of cultures, I suspect that there is significant variability based on how race is conceptualized in a particular culture.

116 critical period before the age of twelve: McKone et al. 2019.

116 white observers were worse: Levin 2000.

7. TURN AND FACE THE STRANGE

119 furniture left behind by the previous tenant: To our surprise, he showed up at our apartment about two years later to pick up that furniture. Fortunately, he changed his mind and let us keep the furniture after seeing its decrepit condition.

119 Dave, who was active in the campus Young Democrats: Whenever I am in Washington, D.C., I enjoy getting together with Dave—who is now a Republican—to get his unique take on the latest political issues.

120 But those times we get it wrong: Den Ouden, Kok, and De Lange 2012; Grossberg 1976; O'Reilly et al. 2012.

120   our eyes are directed by: See, for example, Henderson et al. 2019; Henderson and Hayes 2017; and Hayes and Henderson 2021.

121   a major evolutionary function of the hippocampus: O'Keefe and Nadel 1978.

121   Several subsequent studies have confirmed: See Hannula et al. 2010; Ryan and Shen 2020.

121   hippocampal response to novelty: Endel Tulving and colleagues (Tulving et al. 1994) reported hippocampal activation in response to novel pictures, and this was replicated and extended by Chantal Stern and colleagues (Stern et al. 1996). The fMRI study described here is Wittmann et al. 2007, and the Alzheimer's study is Düzel et al. 2022. For reviews on hippocampal novelty responses, see Ranganath and Rainer 2003; Düzel et al. 2010; and Kafkas and Montaldi 2018.

122   your eyes will quickly be drawn: Voss et al. 2017; Meister and Buffalo 2016. See Ryan and Shen 2020 for reviews of the literature on eye movements, memory, and the hippocampus.

122   the hippocampus is critical for attuning: Ryan et al. 2000.

123   what to expect in the here and now: Voss et al. 2017.

123   Debbie Hannula, a postdoctoral researcher: Debbie, together with Neal Cohen, developed the original paradigm (Hannula et al. 2007), and we adapted it for fMRI (Hannula and Ranganath 2009).

124   I occasionally play in a cover band: For the record, I vehemently protested the misspelling Dogz, but the band is more of an oligarchy than a democracy, so I had little say in the matter. Around the same time, I was constantly arguing about the name of my own band in Davis. We eventually settled on a name that nobody particularly liked. Band names are always contentious.

124   his dogs would salivate: Pavlov 1897. Pavlov's method of pairing an innocuous stimulus with another stimulus that should yield a response is called classical or Pavlovian conditioning. Animal training often relies on "operant conditioning," in which a response is paired with a reward. However, one school of thought holds that operant and classical conditioning are two forms of the same thing (Rescorla and Solomon 1967).

124   Pavlov's most interesting contribution: Pavlov 1924, 1927.

124   The *orienting response*: Sokolov 1963 calls it the "orienting reflex" but would probably have been fine to get credit for naming what we now call the orienting response.

125   There's also a coordinated change: See Ranganath and Rainer 2003 for a review.

125 activate your brain's orienting response: This wave is called the P300, or P3 for short. It is an event-related potential, or ERP, meaning an electrical response that happens about one-third of a second after that dog bark is heard. The P3 was first discovered by Sutton (Sutton et al. 1965), but the version of the experiment that I describe is the "3 stimulus oddball" paradigm in Squires, Squires, and Hillyard 1975. The dog bark would elicit a version of the P300 called the P3a. If this all seems esoteric, I suggest reading a review I wrote with Gregor Rainer (Ranganath and Rainer 2003) as well as the reviews in Polich 2007 and Soltani and Knight 2000.

125 Two of the recurring characters: The orienting response, and the underlying brain areas that give rise to it, are covered by Sokolov 1990; Vinogradova 2001; and Ranganath and Rainer 2003.

125 We know this through recordings: See, for example, Hendrickson, Kimble, and Kimble 1969; Knight 1984, 1996; Stapleton and Halgren 1987; and Paller et al. 1992.

125 Grunwald observed a large electrical spike: Grunwald et al. 1995, 1999.

126 devised a study to figure out: To make the description of our study (Axmacher et al. 2010) readable, I've glossed over a few important points. First, it is not straightforward to make inferences about the relative timing of EEG effects in the patients with electrodes in the nucleus accumbens and those with electrodes in the hippocampus. We did not have the opportunity to test any patients with electrodes in both the hippocampus and the nucleus accumbens. It's reasonable to think that, if we could record from every area of the brain in the same patient, the blip of hippocampal activity would have happened before the burst of activity in the accumbens, but with our results, we can't be 100 percent certain that is the case. Second, it's important to clarify that we could not directly measure dopamine release in this study. The recordings of activity in the nucleus accumbens were changes in electrical signals that could be related to standard neural activity without any dopamine getting into the picture. Ideally, we'd like to measure dopamine levels during the surprising events, but unfortunately there aren't any methods to directly measure the real-time release of dopamine in the human brain. I think that the dopamine explanation is reasonable based on research in animal models, reviewed by Lisman and Grace 2005, and based on an (unfortunately unpublished) fMRI study in which we did the same experiment. In that study, unexpected events also elicited increases in activity around the ventral tegmental area, which is thought to be the major source of dopamine for the nucleus accumbens and many other areas in the brain.

128 Pavlov described the orienting response: See Pavlov 1927.

129 The psychologist George Loewenstein: Loewenstein 1994.

129 Neuroscientists have argued that this drive: Kidd and Hayden 2015.

130 Rewards can sometimes reduce: Murayama et al. 2010.

130 Ben Hayden conducted an experiment: Blanchard, Hayden, and Bromberg-Martin 2015; Wang and Hayden 2019.

131 an innovative fMRI study: Kang et al. 2009.

132 Matthias designed an ingenious experiment: Gruber, Gelman, and Ranganath 2014. See also Murphy et al. 2021; Galli et al. 2018; Fandakova and Gruber 2021; and Stare et al. 2018 for other examples of Matthias's trivia paradigm. Also see Gruber and Ranganath 2019 for a review and our theory to explain these results along with related findings from other labs.

133 a personality trait called *openness to experience*: Tupes and Christal introduced the Big Five or Five Factor Model of personality traits, which included the trait of openness to experience, in a 1961 technical report, which was reprinted for broad dissemination in Tupes and Christal 1992. Silvia and Christensen (2020) point out that research on curiosity tends to focus on the acquisition of knowledge in an academic sense. They argue that we need to adopt a broader view of curiosity, tying it to the literature on openness to experience.

134 never left the confines of their home state: Madeline Buiano, "A New Survey Reveals That One in Six Americans Have Never Left Their Home State," Yahoo! News, November 8, 2021, https://news.yahoo.com/survey-reveals-one-six-americans-154737064.html.

134 Anxiety and curiosity seem like opposites: In rats, the ventral hippocampus is analogous to the front-most (or anterior) part of the hippocampus in humans, and this part is the most extensively interconnected with the amygdala. Jeffrey Gray (1982) reviewed a large body of evidence motivating the theory that the hippocampus plays a central role in anxiety. David Bannerman's lab did pioneering work on the role of the ventral hippocampus in anxiety-like behaviors, and Bannerman et al. 2004 provides a good review of the relevant evidence. Lisman and Grace's (2005) theory about how surprise and novelty influence memory also focused specifically on the ventral/anterior hippocampus.

135 According to Paul Silvia: Silvia 2008.

8. PRESS PLAY AND RECORD

139 Richard Ivens, who had confessed: The section on Richard Ivens was researched from Douglas Starr, "Remembering a Crime That You Didn't Commit," *New Yorker*, March 5, 2015, https://www.newyorker.com/tech/annals-of-technology/false-memory-crime; and Münsterberg 1923.

140 "what I said or did during that time": Romeo Vitelli, "The Problem with

Richard Ivens," *Providentia*, September 22, 2019, https://drvitelli.typepad
.com/providentia/2019/09/the-problem-with-richard-ivens.html.

141 Accessing a memory is more like hitting: In this chapter, I am sidestepping
the thorny question of *how* memories are updated. At one extreme, it
might be the case that, when we recover a past event, the original memory
is erased, replaced by the memory that we construct at the moment of
remembering. At another extreme, one might assume that every time
we remember the same event, we form a new memory, and as a result
we can get mixed up between the memory for the original event and the
memory for the time that we remembered that event (i.e., we can't tell
whether we are remembering an event or remembering that time that
we were remembering the event). This is ultimately a difficult question
to conclusively resolve, but my guess is that the answer lies somewhere
in the middle. I suspect that, every time we remember the same event,
that memory becomes modified. New information can be added to that
memory, and older information can become downregulated, but those
traces of the original event might still remain accessible if we have the
right retrieval cues. In other words, it's complicated.

142 Elizabeth started to think maybe she: Jill Neimark, "The Diva of Dis-
closure, Memory Researcher Elizabeth Loftus," *Psychology Today*, January
1996, 48, https://staff.washington.edu/eloftus/Articles/psytoday.htm;
Ann Marsh and Greta Lorge, "How the Truth Gets Twisted," *Stanford
Magazine*, November/December 2012, https://stanfordmag.org/contents
/how-the-truth-gets-twisted; "Elizabeth Loftus on Experiencing False
Memories of Her Mother's Drowning," *Origins* (podcast), January 13,
2020, https://www.youtube.com/watch?v=WSLRD_qWB4Q.

145 law enforcement agencies across: Noah Caldwell et al., "America's Satanic
Panic Returns—This Time Through QAnon," NPR, May 18, 2021, https://
www.npr.org/2021/05/18/997559036/americas-satanic-panic-returns
-this-time-through-qanon.

146 dubbed the "memory wars": Crews 1995.

146 But her experience with Jenny: Loftus 2005.

146 Working with her research assistant: Loftus and Pickrell 1995.

147 about one in three people can become convinced: This estimate comes
from a pooled analysis of results from eight studies by Scoboria et al. 2017,
and I am limiting the estimate to confident false memories, as opposed to
beliefs about what could have happened. Specifically, Scoboria et al. use the
following criteria: "[1] verbal statements of 'remembering'; [2] acceptance
of suggested information, which indirectly indicates a degree of belief in
the occurrence of the event; [3] elaboration beyond suggested information, which indicates extension of a suggestion beyond acceptance of and

compliance with suggested material; [4] presence and quality of mental imagery; [5] coherence of memory narratives; [6] evidence of emotional experience; and [7] no rejection of the suggested event." That said, this number should be taken with a grain of salt, given that, in any given study, there can be moderator variables (described later) that affect the exact numbers. Arce et al. 2023 provides a more recent and extensive quantitative meta-analysis of results from thirty studies and describes many of these moderator effects, though they use a less stringent standard for the existence of a "rich" false memory. The main point here is not that the one in three estimate is the precise proportion in all situations but rather that studies have consistently found convincing evidence of memory implantation in a significant proportion of subjects.

147 ranging from childhood memories: Loftus and Davis 2006.

148 Guided imagery techniques such as hypnosis: Kloft et al. 2020, 2021; Scoboria et al. 2002; Thomas and Loftus 2002; Arbuthnott, Arbuthnott, and Rossiter 2001.

148 effects of memory implantation can be reversed: Oeberst et al. 2021; Ghetti and Castelli 2006.

148 one individual who recovered memories: Pendergrast 1996.

149 At the end of the experiment: The experiment described here was conducted by Julia Shaw and Stephen Porter (2015), who reported that 70 percent of their participants developed "rich false memories" of having committed the crime. A reanalysis by Kimberley Wade and colleagues (Wade, Garry, and Pezdek 2018) showed that about 30 percent actually remembered committing the crime and the remaining 40 percent didn't remember the event but believed that they had committed a crime and speculated about what it could have been. Wade, Garry, and Pezdek's numbers line up remarkably well with other studies of memory implantation.

149 The "Reid technique": The Reid technique was outlined in Inbau et al. 2001; Saul Kassin (2008), who has extensively researched this topic, recounts numerous disturbing real-life examples of false confessions elicited by the Reid technique. Linda Henkel and Kimberly Coffman (2004) dissect the Reid technique, outlining how it can elicit false memories.

150 Consider the case of Jennifer Thompson: My description of this case is based on interviews reported on an episode of PBS's *Frontline*, titled "What Jennifer Saw," at https://www.pbs.org/wgbh/pages/frontline/shows/dna/etc/script.html. It should be noted that Jennifer Thompson made amends with Ron Cotton, and the two wrote a memoir about the experience. They have given numerous talks to raise awareness about wrongful convictions and to advocate for procedural reforms (particularly in relation to eyewitness memory).

151 Media articles on the malleability of memory: See Freyd 1998 and Roediger and McDermott 1996 for different perspectives on the general relevance of lab research to the "memory wars."

152 under some conditions, misinformation: Stawarczyk et al. 2020.

152 Survivors can be remarkably accurate: Goodman et al. 2003; Alexander et al. 2005; Ghetti et al. 2006. These studies investigated memory performance in young adults who, more than a decade earlier, were involved in criminal prosecutions as victims of sexual abuse. Ghetti et al. 2006, which included results from a sample of 175 interviewees, is most relevant to this point. The authors had details from multiple sources derived from prosecutor files, caregivers, and child victims. Ghetti and her colleagues found that, on average, severity of abuse was positively associated with memory accuracy, despite the fact that they were more likely to have reported past forgetting of the details. The interviews did not indicate that the participants had truly repressed memories in the sense that the memory was completely unavailable at the time of the reported forgetting. The authors note: "When individuals who reported forgetting were asked whether they would have remembered the abuse if asked about it, only 5 responded negatively (i.e., less than 4% of our sample of disclosers). Furthermore, from the descriptions of the recovery triggers, we learned that none of the individuals who provided descriptions of their recovery experiences underwent a long-lasting loss of memory (eventually recovered in adulthood over the course of therapy)." One important caveat here is that the study focused on abuse survivors whose cases were reported and prosecuted. It is possible that other survivors, whose cases were not prosecuted, might have had less reliable memories for the abuse. Also, it is possible that the relationship between trauma severity and accuracy might not apply to people who experienced other forms of trauma. The main point that is worth emphasizing is that survivors of severe childhood trauma can recall details remarkably well, which does not accord with the extreme view that repression is the primary psychological response to childhood trauma.

153 survivor might not have thought: See studies on the "forgot it all along" effect (e.g., Arnold and Lindsay 2002).

153 Perhaps the most interesting aspect of memory updating: Anderson and Hulbert 2021.

154 Karim Nader and Joseph LeDoux published research: Nader, Schafe, and LeDoux 2000. See Riccio, Millin, and Bogart 2006 for a discussion of the history of work in this area.

155 studies have had varying degrees of success: See Schiller and Phelps 2011; Chalkia et al. 2020; Stemerding et al. 2022; Jardine et al. 2022. This might

be the case for a number of reasons. My UC Davis colleague Brian Wilt-gen, who studies memory consolidation in mice, believes that the mixed results reflect that reconsolidation can only be disrupted if the memory is actually recalled. For instance, people with PTSD would have to vividly recall their traumatic experience before taking a drug that would block the fear response. If, instead, they were simply reminded of the event and kept themselves from reliving the traumatic memory, then the memory would not adequately be disrupted. This explanation fits with behavioral research in humans by Hupbach et al. 2008 showing that reminders can lead memories to be updated when people are put back into the same context in which the original event took place.

Sam Gershman, one of my collaborators at Harvard University, has a related idea—that even if someone is reminded of a traumatic event, and a drug is used to suppress the fear response, reconsolidation might still not be disrupted unless you interpret the effect of the drug the right way (Gershman et al. 2017). If you experienced the effect of the drug as an indicator that the memory has become less threatening, then the memory could be disrupted (see also Sinclair and Barense 2019). If, on the other hand, you (more accurately) conclude that the drug is merely suppressing your fear response, you might not unlearn the fear associated with the traumatic event.

155 number of clinical trials of reconsolidation-based treatments: Astill Wright et al. 2021.

155 effectiveness of psychedelic-assisted psychotherapy: Feduccia and Mithoefer 2018.

9. SOME PAIN, MORE GAIN

160 Error-driven learning is a well-established principle in the brain's motor system: See, for example, McDougle, Ivry, and Taylor 2016.

160 Memory researchers have long known: Robert and Elizabeth Bjork have used the umbrella term *desirable difficulties* to describe this phenomenon (e.g., Bjork and Bjork 2011).

160 What if, instead of studying, you trained: Research on the testing effect goes back at least to the work of Edwina Abbott (1909). In the text, I emphasized the SSSS and SSST conditions of the experiment in Roediger and Karpicke 2006. See Karpicke and Roediger 2008 for a particularly compelling demonstration that, even when initial learning is equated, testing can dramatically improve retention.

161 Roddy once wrote that: Roediger 2008.

161 The testing effect has been shown: Roediger and Butler 2011 provides an

accessible review of the literature on the testing effect, and Rowland 2014's meta-analysis substantiates the reliability of the effect. Note that I emphasize error-driven learning in this chapter, but numerous mechanisms probably contribute to the testing effect. For more on the testing effect and also the spacing effect (discussed later), check out *Make It Stick*, an easy-to-read book that's full of practical tips from Roddy and his Washington University colleague Mark McDaniel (Brown, Roediger, and McDaniel 2014).

162 "pre-testing" turns out to be remarkably effective: See, e.g., Richland, Kornell, and Kao 2009; Potts, Davies, and Shanks 2019.

162 an intriguing theory to explain this principle of memory: Carrier and Pashler 1992.

162 modern artificial intelligence systems learn: For an overall review of how error-driven learning works in neural networks, see O'Reilly et al. 2012. In case you are curious about how much struggle is good, Wilson et al. 2019 has detailed analyses suggesting that, at least in machine learning, optimal learning happens with a 15 percent error rate. It's not clear if this applies to humans under all circumstances, but if you're making a mistake just over one out of ten tries, that is probably the sweet spot for learning while still maintaining your self-esteem.

162 simulated the testing effect: The first implementation of error-driven learning (EDL) in this model was reported by Ketz, Morkonda, and O'Reilly 2013. Zheng et al. 2022 expanded the implementation of EDL in the Theremin (Total Hippocampal ERror MINimization) model and showed that the testing effect could only be simulated with EDL implemented. Liu, O'Reilly, and Ranganath 2021 reviews the broader literature on the testing effect and explains how EDL can explain the testing effect. Note that, in this model, "error" reflects the degree of mismatch between what the hippocampus is trying to encode and what the hippocampus should be encoding. Ketz, Morkonda, and O'Reilly 2013 proposed that EDL happens during theta oscillations, which alternate hippocampal output with input providing the "ground truth" feedback about the target memory. Theta oscillations are not necessary, however, as one might see such dynamics even during nonrhythmic up-and-down states. Also note that even if the hippocampus encodes a fairly close match to what should have been encoded, there will always be some error during retrieval, which provides an opportunity for the model to learn how to do a better job next time.

164 recalling one memory can make it harder: The standard RIF paradigm was first reported by Anderson, Bjork, and Bjork 1994. The literature on this topic is massive. See Hauer and Wessel 2006 for an example of RIF in autobiographical memory. Jonker, Seli, and MacLeod 2013 and Murayama

et al. 2014 review some of the literature on this topic from different theoretical perspectives.

164 When people were tested on one: Chan, McDermott, and Roediger 2006. See also Chan 2009; Jonker et al. 2018; Liu and Ranganath 2021; and Oliva and Storm 2022 for a meta-analysis of this literature.

165 Error-driven learning strengthens the memory: Norman, Newman, and Detre 2007 introduced a computational model that focuses specifically on this mechanism.

166 get much more bang for your buck: Ebbinghaus (1885) established the benefit of spaced learning (aka distributed practice) in his studies on retention of trigrams. See Carpenter et al. 2012 for a review and practical applications of research on the spacing effect.

166 James Antony, another scientist in my lab: See Antony et al. 2022.

166 the benefit you get from spacing out your learning: Pashler et al. 2009's "multiscale context model" showed how changes in context between study events, along with error-driven learning, could account for spacing effects. James Antony (2022) implemented a similar mechanism in his HipSTeR (Hippocampus with Spectral Temporal Representations) model, which emulates the computational architecture of the hippocampus. In working on these topics, I have learned that the success of a computational model is proportional to the catchiness and cleverness of its name.

167 update those memories until they have: Some memory researchers would say that, when this happens, the memory becomes "semanticized," because it is no longer tied to a single episode, but I think there is a difference between a decontextualized episodic memory and a semantic memory. Decontextualization could lead you to have a blurry memory of a visit to Paris, but that is not the same as having a semantic memory that tells you about the kinds of things you might see in French cities.

167 our brains are hard at work during sleep: For an accessible review of what happens in the brain during sleep, I recommend Matt Walker's wonderful 2017 book, *Why We Sleep*.

167 Some of that work is housekeeping: Xie et al. 2013.

168 *slow-wave sleep* (SWS): See Singh, Norman, and Schapiro 2022.

168 a wonderfully orchestrated interaction: Staresina et al. 2015 provides a great demonstration of these oscillations in the human brain. See Geva-Sagiv and Nir 2019 and Navarrete, Valderrama, and Lewis 2020 for reviews of SWS oscillations, and Kaplan et al. 2016 for evidence of DMN activation during hippocampal ripples.

168 Dreams occur during REM sleep: Wamsley and Stickgold 2011; Zadra and Stickgold 2021.

168 Many in the neuroscience community have proposed: The most popular variant is the Active Systems Consolidation theory (see, e.g., Diekelmann and Born 2010), which is based on standard systems consolidation theory. I favor alternative theories that emphasize the dynamic and interactive aspects of memory (Moscovitch et al. 2016; Yonelinas et al. 2019).

169 Xiaonan Liu ran a number of studies: Liu and Ranganath 2021.

169 Xiaonan's simulations suggest that: These simulations were conducted with Xiaonan's TEACH (TEsting Activated Cortico-Hippocampal interaction) model, Liu, Ranganath, and O'Reilly 2022. I got the phrase "tapestry of knowledge" from conversations with Matt Walker, who is a gifted wordsmith.

170 sleep helps us integrate what we have: Lewis and Durrant 2011.

170 Revolutionary research by UC Irvine psychologist: Mednick, Nakayama, and Stickgold 2003. See also Mednick's book *Take a Nap! Change Your Life* (Mednick and Ehrman 2006).

171 are also evident when a rat: See Joo and Frank 2018 for a review.

171 areas of the brain that were engaged: See, e.g., Tambini, Ketz, and Davachi 2010.

171 the hippocampus reactivates memories: Gruber et al. 2016.

171 During these down states: See Mednick 2020 for a readable book on this topic filled with actionable science-backed tips.

172 we might be able to hack into sleep: Ken was inspired by Rasch et al. 2007, which used odors to reactivate memories during sleep.

172 To test his unorthodox theory: Rudoy et al. 2009. I say that targeted memory reactivation (TMR) "strengthened" the memory as a simplification. Technically, TMR reduced the typical forgetting that you see over time, and we can't be sure that any memories were "strengthened" per se.

172 Ken's team has explored numerous ways: This section refers to Antony et al. 2012; Batterink and Paller 2017; Hu et al. 2015; and Sanders et al. 2019. See also Hu et al. 2020 and Paller, Creery, and Schechtman 2021.

172 Simona Ghetti has used a similar technique: Prabhakar et al. 2018 is described in the paragraph. See also Johnson et al. 2021 and Mooney et al. 2021.

10. WHEN WE REMEMBER TOGETHER

175 40 percent of our conversational time: Eggins and Slade 2004.

176 French philosopher Maurice Halbwachs: Halbwachs 1992.

177 construct a *life narrative*: Most of this section is influenced by the work of Robyn Fivush and colleagues (Nelson and Fivush 2004; Fivush 2008). Dan

McAdams has written about the role of autobiographical memory "life stories" in our sense of self (e.g., McAdams 2008).

177 mother-child interactions can have a formative influence: Fivush 2004; Nelson and Fivush 2004.

180 We tailor narratives for our audience: To learn more about the effects of audience tuning on one's memory, see studies of the "saying is believing" effect by E. Tory Higgins and colleagues (e.g., Higgins and Rholes 1978; Echterhoff et al. 2008).

180 The transformative effect of social interactions: Therapy methods that are based on different theories and use different approaches can often show equivalent effects on treatment outcomes, suggesting that some benefits of psychotherapy might come from elements that are not specific to any single theory. I suspect that a key ingredient of most effective psychotherapy techniques is the process of sharing and updating difficult memories in the context of a trusting, supportive relationship.

180 when people hear part of a story: To test this hypothesis, my student Brendan Cohn-Sheehy wrote several short stories with side plots that had a recurring character, and we looked at brain activity when people had to put the pieces together from the various side plots. We called it the "Kramer experiment," after the character from the TV show *Seinfeld*, who would periodically walk into Jerry's apartment to talk about his latest experiences. To understand the Kramer subplots, viewers would have to periodically recall past information and integrate it with the current scene in order to put together the full narrative for his character in that episode. See Cohn-Sheehy et al. 2021, 2022.

181 two research groups published papers that challenged: This experimental approach was reported in the back-to-back papers Weldon and Bellinger 1997 and Basden et al 1997.

181 those working in a group had worse: My synopsis of the research on collaborative inhibition is based on an excellent review of this literature by Suparna Rajaram (in press).

181 Recalling in groups can also have a homogenizing effect: See Greeley and Rajaram 2023 and Greeley et al. 2023.

181 recollections of groups disproportionately reflect: Cuc et al. 2006; see Hirst 2010 for a review.

182 Suparna Rajaram, professor of psychology: Choi et al. 2014; Greeley et al. 2023; Luhmann and Rajaram 2015.

182 *Collaborative facilitation*, as this is known: Meade, Nokes, and Morrow 2009, for instance, reported on the effects of collaborative facilitation on memory with those with expertise in aviation.

182   psychologist Celia Harris at Macquarie University: Celia Harris and col-
      leagues have documented when and how couples show collaborative facili-
      tation (Harris et al. 2011, 2014; Barnier et al. 2018).

183   Some findings even suggest: Dixon 2011; Gerstorf et al. 2009; and Rauers
      et al. 2011 provide evidence for how collaboration between members of
      a couple can improve everyday memory in older adults.

183   people with memory disorders: There is a surprisingly deep literature on
      the relationship between memory and common ground in language—that
      is, how memory helps us achieve a shared understanding with our conver-
      sational partners, and how a shared understanding can, in turn, compen-
      sate for memory deficits. Melissa Duff and Neal Cohen have extensively
      studied this area with an innovative combination of neuroscience and
      linguistic analysis, including studies of patients with hippocampal amnesia
      (Duff et al. 2006, 2008; Rubin et al. 2011; Brown-Schmidt and Duff 2016).
      See also Horton and Gehrig 2005, 2016.

184   he ran an experiment on "serial reproduction": Bartlett 1932.

185   Memory researchers have since used Bartlett's: Kashima 2000; Lyons and
      Kashima 2001, 2003; Choi, Kensinger, and Rajaram 2017.

185   negativity bias in social transmission of memories: Bebbington et al. 2017.

185   extensively studied memory distortion in individuals: My review of social
      contagion draws upon the original studies Betz, Skowronski, and Ostrom
      1996; Roediger, Meade, and Bergman 2001; and Meade and Roediger 2002;
      as well as the comprehensive and thoughtful review paper Maswood and
      Rajaram 2019.

186   especially effective at spreading misinformation: Cuc et al. 2006; Peker and
      Tekcan 2009; Koppel et al. 2014.

186   protect us from social contagion: Andrews and Rapp 2014; Maswood,
      Luhmann, and Rajaram 2022.

187   Loftus's team ran an online experiment: Frenda et al. 2013. See also Wil-
      liam Saletan, "The Memory Doctor," *Slate*, June 4, 2010, http://www
      .slate.com/articles/health_and_science/the_memory_doctor/2010/06
      /the_memory_doctor.html.

187   What makes people susceptible to fake news?: This synopsis draws on
      the recent reviews by Schacter 2022 and Pennycook and Rand 2021. Inter-
      estingly, Pennycook and Rand 2021 mentions a related line of work by
      Pennycook on people's willingness to believe "pseudo-profound bullshit"
      (see, e.g., Pennycook et al. 2015; Pennycook and Rand 2020). Sadly, I could
      not find a way to work it into this chapter, but I encourage the interested
      reader to check it out.

188   repeatedly exposed to false information: The effect of familiarity on belief,
      now known as the *illusory truth effect* when applied to false information,

was reported in Hasher, Goldstein, and Toppino 1977 and elaborated on since in a number of studies. Unkelbach et al. 2019 provides a nice review of this literature.

188 Recent research has highlighted the use of "push polls,": Saletan 2000 describes the misleading questions used in push polls from the Bush campaign. See also Gooding 2004. The study on push polls and false memories is Murphy et al. 2021.

189 over five thousand participants read through headlines: The experiment investigating the effect of fact-checking on memory for fake news was reported in Brashier et al. 2021.

CODA: DYNAMIC MEMORIES

192 memory has a distinct role: My perspective on memory changes across the life span was influenced by an interview with Alison Gopnik on the *Stanford Psychology Podcast* (Cao, n.d.) and, in a more abstract sense, by the Hindu conceptualization of development as a progression through different life stages, each with a different purpose.

192 Alison Gopnik has argued: Gopnik 2020.

193 The persistence of semantic memory: See, for example, Grady 2012 for a review.

193 elders play a central role: Viscogliosi et al. 2020.

193 experience accumulated over a lifetime by orca grandmothers: I admittedly went down a rabbit hole for this section because orcas are fascinating animals. Rendell and Whitehead 2001 discusses research on cultural transmission in killer whales and cetaceans in general, and this is covered in a more accessible way in Stiffler 2011. Simulations from Nattrass et al. 2019 show how the extended postmenopausal period of orcas might improve evolutionary fitness.

BIBLIOGRAPHY

Abbott, E. E. 1909. "On the Analysis of the Factors of Recall in the Learning Process." *Psychological Monographs* 11:159–77.

Abernathy, K., L. J. Chandler, and J. J. Woodward. 2010. "Alcohol and the Prefrontal Cortex." *International Review of Neurobiology* 91:289–320.

Addis, Donna Rose, Alana T. Wong, and Daniel L. Schacter. 2007. "Remembering the Past and Imagining the Future: Common and Distinct Neural Substrates During Event Construction and Elaboration." *Neuropsychologia* 45 (7): 1363–77.

Adler, Orly, and Ainat Pansky. 2020. "A 'Rosy View' of the Past: Positive Memory Biases." Chap. 7 in *Cognitive Biases in Health and Psychiatric Disorders*, edited by Tatjana Aue and Hadas Okon-Singer, 139–71. New York: Academic Press.

Adolphs, Ralph, Larry Cahill, Rina Schul, and Ralf Babinsky. 1997. "Impaired Declarative Memory for Emotional Material Following Bilateral Amygdala Damage in Humans." *Learning & Memory* 4 (3): 291–300.

Adolphs, Ralph, L. Mlodinow, and L. F. Barrett. 2019. "What Is an Emotion?" *Current Biology* 29 (20): R1060–R1064.

Aggleton, J. P., and M. W. Brown. 1999. "Episodic Memory, Amnesia, and the Hippocampal-Anterior Thalamic Axis." *Behavioral and Brain Sciences* 22 (3): 425–44.

Ahmed, S. H. 2010. "Validation Crisis in Animal Models of Drug Addiction: Beyond Non-disordered Drug Use Toward Drug Addiction." *Neuroscience & Biobehavioral Reviews* 35 (2): 172–84.

Alexander, Kristen Weede, Jodi A. Quas, Gail S. Goodman, Simona Ghetti,

Robin S. Edelstein, Allison D. Redlich, Ingrid M. Cordon, and David P. H. Jones. 2005. "Traumatic Impact Predicts Long-Term Memory for Documented Child Sexual Abuse." *Psychological Science* 16 (1): 33–40.

Alexander, M. P., D. Stuss, and S. Gillingham. 2009. "Impaired List Learning Is Not a General Property of Frontal Lesions." *Journal of Cognitive Neuroscience* 21 (7): 1422–34.

Amer, Tarek, Karen L. Campbell, and Lynn Hasher. 2016. "Cognitive Control as a Double-Edged Sword." *Trends in Cognitive Sciences* 20 (12): 905–15.

Anderson, M. C., R. A. Bjork, and E. L. Bjork. 1994. "Remembering Can Cause Forgetting: Retrieval Dynamics in Long-Term Memory." *Journal of Experimental Psychology: Learning, Memory, and Cognition* 20 (5): 1063.

Anderson, M. C., and J. C. Hulbert. 2021. "Active Forgetting: Adaptation of Memory by Prefrontal Control." *Annual Review of Psychology* 72 (January 4): 1–36. https://doi.org/10.1146/annurev-psych-072720-094140.

Anderson, Richard C., and James W. Pichert. 1978. "Recall of Previously Unrecallable Information Following a Shift in Perspective." *Journal of Verbal Learning and Verbal Behavior* 17 (1): 1–12.

Andrews, Jessica J., and David N. Rapp. 2014. "Partner Characteristics and Social Contagion: Does Group Composition Matter?" *Applied Cognitive Psychology* 28 (4): 505–17.

Annese, J., N. M. Schenker-Ahmed, H. Bartsch, P. Maechler, C. Sheh, N. Thomas, J. Kayano, et al. 2014. "Postmortem Examination of Patient HM's Brain Based on Histological Sectioning and Digital 3D Reconstruction." *Nature Communications* 5 (1): 3122.

Anspach, Carolyn Kiser. 1934. "Medical Dissertation on Nostalgia by Johannes Hofer, 1688." *Bulletin of the Institute of the History of Medicine* 2 (6): 376–91.

Antony, J., E. W. Gobel, J. K. O'Hare, P. J. Reber, and K. A. Paller. 2012. "Cued Memory Reactivation During Sleep Influences Skill Learning." *Nature Neuroscience* 15 (8): 1114–16.

Antony, J., X. L. Liu, Y. Zheng, C. Ranganath, and R. C. O'Reilly. 2024. "Memory out of Context: Spacing Effects and Decontextualization in a Computational Model of the Medial Temporal Lobe." *Psychological Review*, https://doi.org/10.1037/rev0000488.

Arbuthnott, Katherine D., Dennis W. Arbuthnott, and Lucille Rossiter. 2001. "Guided Imagery and Memory: Implications for Psychotherapists." *Journal of Counseling Psychology* 48 (2): 123.

Arce, Ramón, Adriana Selaya, Jéssica Sanmarco, and Francisca Fariña. 2023. "Implanting Rich Autobiographical False Memories: Meta-analysis for Forensic Practice and Judicial Judgment Making." *International Journal of Clinical and Health Psychology* 23 (4): 100386.

Arnold, M. M., and D. S. Lindsay. 2002. "Remembering Remembering." *Journal of Experimental Psychology: Learning, Memory, and Cognition* 28 (3): 521–29. https://doi.org/10.1037/0278-7393.28.3.521.

Arnsten, Amy F. T. 2009a. "The Emerging Neurobiology of Attention Deficit Hyperactivity Disorder: The Key Role of the Prefrontal Association Cortex." *Journal of Pediatrics* 154 (5): I–S43.

———. 2009b. "Stress Signaling Pathways That Impair Prefrontal Cortex Structure and Function." *Nature Reviews Neuroscience* 10 (6): 410–22.

Avery, Michael C., and Jeffrey L. Krichmar. 2017. "Neuromodulatory Systems and Their Interactions: A Review of Models, Theories, and Experiments." *Frontiers in Neural Circuits* 11:108.

Axmacher, Nikolai, Michael X. Cohen, Juergen Fell, Sven Haupt, Matthias Dümpelmann, Christian E. Elger, Thomas E. Schlaepfer, Doris Lenartz, Volker Sturm, and Charan Ranganath. 2010. "Intracranial EEG Correlates of Expectancy and Memory Formation in the Human Hippocampus and Nucleus Accumbens." *Neuron* 65 (4): 541–49.

Baddeley, Alan, and Graham Hitch. 1974. "Working Memory." In *The Psychology of Learning and Motivation: Advances in Research and Theory*, edited by G. H. Bower, 8:47–89. New York: Academic Press.

Baddeley, Alan, and Barbara Wilson. 1988. "Frontal Amnesia and the Dysexecutive Syndrome." *Brain and Cognition* 7 (2): 212–30.

Badre, D. 2020. *On Task: How Our Brain Gets Things Done*. Princeton, NJ: Princeton University Press.

Baird, Amee, Olivia Brancatisano, Rebecca Gelding, and William Forde Thompson. 2018. "Characterization of Music and Photograph Evoked Autobiographical Memories in People with Alzheimer's Disease." *Journal of Alzheimer's Disease* 66 (2): 693–706.

Baldassano, C., J. Chen, A. Zadbood, J. W. Pillow, U. Hasson, and K. A. Norman. 2017. "Discovering Event Structure in Continuous Narrative Perception and Memory." *Neuron* 95 (3): 709–21.

Baldassano, C., U. Hasson, and K. A. Norman. 2018. "Representation of Real-World Event Schemas During Narrative Perception." *Journal of Neuroscience* 38 (45): 9689–99.

Bannerman, D. M., J. N. P. Rawlins, S. B. McHugh, R. M. J. Deacon, B. K. Yee, T. Bast, W.-N. Zhang, H. H. J. Pothuizen, and J. Feldon. 2004. "Regional Dissociations Within the Hippocampus—Memory and Anxiety." *Neuroscience & Biobehavioral Reviews* 28 (3): 273–83.

Bar, Moshe. 2022. *Mindwandering*. New York: Hachette.

Barasch, A., K. Diehl, J. Silverman, and G. Zauberman. 2017. "Photographic Memory: The Effects of Volitional Photo Taking on Memory for Visual and Auditory Aspects of an Experience." *Psychological Science* 28 (8) (August): 1056–66. https://doi.org/10.1177/0956797617694868.

Barnett, A. J., M. Nguyen, J. Spargo, R. Yadav, B. Cohn-Sheehy, and C. Ranganath. In press. "Hippocampal-Cortical Interactions During Event Boundaries Support Retention of Complex Narrative Events." *Neuron*.

Barnett, Alexander J., Walter Reilly, Halle R. Dimsdale-Zucker, Eda Mizrak, Zachariah Reagh, and Charan Ranganath. 2021. "Intrinsic Connectivity Reveals Functionally Distinct Cortico-hippocampal Networks in the Human Brain." *PLoS Biology* 19 (6): e3001275.

Barnier, A. J., C. B. Harris, T. Morris, and G. Savage. 2018. "Collaborative Facilitation in Older Couples: Successful Joint Remembering Across Memory Tasks." *Frontiers in Psychology* 9:2385.

Bartlett, Frederic Charles. 1928a. *Psychology and Primitive Culture*. CUP Archive.

———. 1928b. "Types of Imagination." *Philosophy* 3 (9): 78–85.

———. 1932. *Remembering: A Study in Experimental and Social Psychology*. Cambridge: Cambridge University Press.

———. 2014. *Psychology and the Soldier*. Cambridge: Cambridge University Press.

Basden, B. H., D. R. Basden, S. Bryner, and R. L. Thomas III. 1997. "A Comparison of Group and Individual Remembering: Does Collaboration Disrupt Retrieval Strategies?" *Journal of Experimental Psychology: Learning, Memory, and Cognition* 23 (5): 1176.

Batterink, L. J., and K. A. Paller. 2017. "Sleep-Based Memory Processing Facilitates Grammatical Generalization: Evidence from Targeted Memory Reactivation." *Brain and Language* 167:83–93.

Bauer, P. 2004. "Oh Where, Oh Where Have Those Early Memories Gone? A Developmental Perspective on Childhood Amnesia." *Psychological Science Agenda* 18:12.

Beaty, Roger E., Preston P. Thakral, Kevin P. Madore, Mathias Benedek, and Daniel L. Schacter. 2018. "Core Network Contributions to Remembering the Past, Imagining the Future, and Thinking Creatively." *Journal of Cognitive Neuroscience* 30 (12): 1939–51.

Bebbington, Keely, Colin MacLeod, T. Mark Ellison, and Nicolas Fay. 2017. "The Sky Is Falling: Evidence of a Negativity Bias in the Social Transmission of Information." *Evolution and Human Behavior* 38 (1): 92–101.

Bechara, Antoine, Daniel Tranel, Hanna Damasio, Ralph Adolphs, Charles Rockland, and Antonio R. Damasio. 1995. "Double Dissociation of Conditioning and Declarative Knowledge Relative to the Amygdala and Hippocampus in Humans." *Science* 269 (5227): 1115–18.

Becker, Jacqueline H., Jenny J. Lin, Molly Doernberg, Kimberly Stone, Allison Navis, Joanne R. Festa, and Juan P. Wisnivesky. 2021. "Assessment of Cognitive Function in Patients After COVID-19 Infection." *JAMA Network Open* 4 (10): e2130645.

Ben-Yakov, A., N. Eshel, and Y. Dudai. 2013. "Hippocampal Immediate Post-stimulus Activity in the Encoding of Consecutive Naturalistic Episodes." *Journal of Experimental Psychology: General* 142 (4): 1255.

Ben-Yakov, A., and R. N. Henson. 2018. "The Hippocampal Film Editor: Sensitivity and Specificity to Event Boundaries in Continuous Experience." *Journal of Neuroscience* 38 (47): 10057–68.

Berridge, Kent C., and Terry E. Robinson. 2016. "Liking, Wanting, and the Incentive-Sensitization Theory of Addiction." *American Psychologist* 71 (8): 670.

Betz, Andrew L., John J. Skowronski, and Thomas M. Ostrom. 1996. "Shared Realities: Social Influence and Stimulus Memory." *Social Cognition* 14 (2): 113.

Beversdorf, David Q., Brian W. Smith, Gregory P. Crucian, Jeffrey M. Anderson, Jocelyn M. Keillor, Anna M. Barrett, John D. Hughes, et al. 2000. "Increased Discrimination of 'False Memories' in Autism Spectrum Disorder." *Proceedings of the National Academy of Sciences* 97 (15): 8734–37.

Bjork, E. L., and R. A. Bjork. 2011. "Making Things Hard on Yourself, but in a Good Way: Creating Desirable Difficulties to Enhance Learning." *Psychology and the Real World: Essays Illustrating Fundamental Contributions to Society* 2:56–64.

Blanchard, T. C., B. Y. Hayden, and E. S. Bromberg-Martin. 2015. "Orbitofrontal Cortex Uses Distinct Codes for Different Choice Attributes in Decisions Motivated by Curiosity." *Neuron* 85 (3): 602–14.

Blumenfeld, Robert S., and Charan Ranganath. 2019. "The Lateral Prefrontal Cortex and Human Long-Term Memory." *Handbook of Clinical Neurology* 163:221–35.

Bower, Gordon H. 1981. "Mood and Memory." *American Psychologist* 36 (2): 129.

Brashier, N. M., G. Pennycook, A. J. Berinsky, and D. G. Rand. 2021. "Timing Matters When Correcting Fake News." *Proceedings of the National Academy of Sciences* 118 (5): e2020043118.

Braver, T. S., J. D. Cohen, L. E. Nystrom, J. Jonides, E. E. Smith, and D. E. Noll. 1997. "A Parametric Study of Prefrontal Cortex Involvement in Human Working Memory." *Neuroimage* 5 (1): 49–62.

Brown, A. S. 1991. "A Review of the Tip-of-the-Tongue Experience." *Psychological Bulletin* 109 (2): 204.

Brown, A. S., and D. R. Murphy. 1989. "Cryptomnesia: Delineating Inadvertent Plagiarism." *Journal of Experimental Psychology: Learning, Memory, and Cognition* 15 (3): 432.

Brown, M. W., and J. Z. Xiang. 1998. "Recognition Memory: Neuronal Substrates of the Judgment of Prior Occurrence." *Progress in Neurobiology* 55 (2): 149–89.

Brown, P. C., H. L. Roediger III, and M. A. McDaniel. 2014. "Make It Stick: The Science of Successful Learning." Cambridge, MA: Harvard University Press.

Brown, R., and D. McNeill. 1966. "The 'Tip of the Tongue' Phenomenon." *Journal of Verbal Learning and Verbal Behavior* 5 (4): 325–37.

Brown-Schmidt, S., and M. C. Duff. 2016. "Memory and Common Ground Processes in Language Use." *Topics in Cognitive Science* 8 (4) (October): 722–36. https://doi.org/10.1111/tops.12224.

Buckner, R. L., J. R. Andrews-Hanna, and D. L. Schacter. 2008. "The Brain's Default Network: Anatomy, Function, and Relevance to Disease." *Annals of the New York Academy of Sciences* 1124 (1): 1–38.

Buda, M., A. Fornito, Z. M. Bergström, and J. S. Simons. 2011. "A Specific Brain Structural Basis for Individual Differences in Reality Monitoring." *Journal of Neuroscience* 31 (40): 14308–13.

Budson, A. E., and E. A. Kensinger. 2023. *Why We Forget and How to Remember Better: The Science Behind Memory.* Oxford: Oxford University Press.

Bukach, C. M., I. Gauthier, and M. J. Tarr. 2006. "Beyond Faces and Modularity: The Power of an Expertise Framework." *Trends in Cognitive Sciences* 10 (4): 159–66.

Buzan, Tony. 1984. *Use Your Perfect Memory.* New York: Dutton.

Cabeza, R., S. M. Rao, A. D. Wagner, A. R. Mayer, and D. L. Schacter. 2001. "Can Medial Temporal Lobe Regions Distinguish True from False? An Event-Related Functional MRI Study of Veridical and Illusory Recognition Memory." *Proceedings of the National Academy of Sciences* 98 (8): 4805–10.

Campbell, Karen L., Lynn Hasher, and Ruthann C. Thomas. 2010. "Hyperbinding: A Unique Age Effect." *Psychological Science* 21 (3): 399–405.

Cao, A. n.d. "Alison Gopnik: How Can Understanding Childhood Help Us Build Better AI?" *Stanford Psychology Podcast*, episode 14. https://podcasts.apple.com/us/podcast/14-alison-gopnik-how-can-understanding-childhood-help/id1574802514?i=1000537173346.

Carpenter, Gail A., and Stephen Grossberg. 1988. "The ART of Adaptive Pattern Recognition by a Self-Organizing Neural Network." *Computer* 21 (3): 77–88.

Carpenter, S. K., N. J. Cepeda, D. Rohrer, S. H. Kang, and H. Pashler. 2012. "Using Spacing to Enhance Diverse Forms of Learning: Review of Recent Research and Implications for Instruction." *Educational Psychology Review* 24: 369–78.

Carrier, Mark, and Harold Pashler. 1992. "The Influence of Retrieval on Retention." *Memory & Cognition* 20: 633–42.

Chalkia, A., N. Schroyens, L. Leng, N. Vanhasbroeck, A. K. Zenses, L. Van Oudenhove, and T. Beckers. 2020. "No Persistent Attenuation of Fear Memories in Humans: A Registered Replication of the Reactivation-Extinction Effect." *Cortex* 129:496–509.

Chan, Jason C. K. 2009. "When Does Retrieval Induce Forgetting and When Does It Induce Facilitation? Implications for Retrieval Inhibition, Testing Effect, and Text Processing." *Journal of Memory and Language* 61 (1): 153–70.

Chan, Jason C. K., K. B. McDermott, and H. L. Roediger III. 2006. "Retrieval-

Induced Facilitation: Initially Nontested Material Can Benefit from Prior Testing of Related Material." *Journal of Experimental Psychology: General* 135 (4): 553.

Chase, William G., and Herbert A. Simon. 1973. "Perception in Chess." *Cognitive Psychology* 4 (1): 55–81.

Chen, J., Y. C. Leong, C. J. Honey, C. H. Yong, K. A. Norman, and U. Hasson. 2017. "Shared Memories Reveal Shared Structure in Neural Activity Across Individuals." *Nature Neuroscience* 20 (1): 115–25.

Choi, H.-Y., H. M. Blumen, A. R. Congleton, and S. Rajaram. 2014. "The Role of Group Configuration in the Social Transmission of Memory: Evidence from Identical and Reconfigured Groups." *Journal of Cognitive Psychology* 26 (1): 65–80.

Choi, Hae-Yoon, Elizabeth A. Kensinger, and Suparna Rajaram. 2017. "Mnemonic Transmission, Social Contagion, and Emergence of Collective Memory: Influence of Emotional Valence, Group Structure, and Information Distribution." *Journal of Experimental Psychology: General* 146 (9): 1247.

Cohen, J. D., W. M. Perlstein, T. S. Braver, L. E. Nystrom, D. C. Noll, J. Jonides, and E. E. Smith. 1997. "Temporal Dynamics of Brain Activation During a Working Memory Task." *Nature* 386 (6625): 604–8.

Cohen, Michael X. 2007. "Individual Differences and the Neural Representations of Reward Expectation and Reward Prediction Error." *Social Cognitive and Affective Neuroscience* 2 (1) (March): 20–30. https://doi.org/10.1093/scan/nsl021.

Cohen, Michael X., Christian E. Elger, and Charan Ranganath. 2007. "Reward Expectation Modulates Feedback-Related Negativity and EEG Spectra." *Neuroimage* 35 (2) (April 1): 968–78. https://doi.org/10.1016/j.neuroimage.2006.11.056.

Cohen, Michael X., and Charan Ranganath. 2005. "Behavioral and Neural Predictors of Upcoming Decisions." *Cognitive, Affective & Behavioral Neuroscience* 5 (2) (June): 117–26. https://doi.org/10.3758/cabn.5.2.117.

———. 2007. "Reinforcement Learning Signals Predict Future Decisions." *Journal of Neuroscience* 27 (2) (January 10): 371–78. https://doi.org/10.1523/JNEUROSCI.4421-06.2007.

Cohen, M. X., J. Young, J. M. Baek, C. Kessler, and C. Ranganath. 2005. "Individual Differences in Extraversion and Dopamine Genetics Predict Neural Reward Responses." *Brain Research: Cognitive Brain Research* 25 (3) (December): 851–61. https://doi.org/10.1016/j.cogbrainres.2005.09.018.

Cohen, Neal J., and Howard Eichenbaum. 1995. *Memory, Amnesia, and the Hippocampal System*. Cambridge, MA: MIT Press.

Cohn-Sheehy, B. I., A. I. Delarazan, J. E. Crivelli-Decker, Z. M. Reagh, N. S. Mundada, A. P. Yonelinas, J. M. Zacks, and C. Ranganath. 2022. "Narratives

Bridge the Divide Between Distant Events in Episodic Memory." *Memory & Cognition* 50 (3): 478–94.

Cohn-Sheehy, B. I., A. I. Delarazan, Z. M. Reagh, J. E. Crivelli-Decker, K. Kim, A. J. Barnett, J. M. Zacks, and C. Ranganath. 2021. "The Hippocampus Constructs Narrative Memories Across Distant Events." *Current Biology* 31 (22): 4935–45.

Cook, Peter, Ashley Prichard, Mark Spivak, and Gregory S. Berns. 2018. "Jealousy in Dogs? Evidence from Brain Imaging." *Animal Sentience* 3 (22): 1.

Cook, Peter F., Colleen Reichmuth, Andrew A. Rouse, Laura A. Libby, Sophie E. Dennison, Owen T. Carmichael, Kris T. Kruse-Elliott, et al. 2015. "Algal Toxin Impairs Sea Lion Memory and Hippocampal Connectivity, with Implications for Strandings." *Science* 350 (6267): 1545–47.

Corkin, S. 2013. *Permanent Present Tense: The Unforgettable Life of the Amnesic Patient, H.M.* New York: Basic Books.

Corkin, S., D. G. Amaral, R. G. González, K. A. Johnson, and B. T. Hyman. 1997. "HM's Medial Temporal Lobe Lesion: Findings from Magnetic Resonance Imaging." *Journal of Neuroscience* 17 (10): 3964–79.

Courtney, S. M., L. Petit, J. M. Maisog, L. G. Ungerleider, and J. V. Haxby. 1998. "An Area Specialized for Spatial Working Memory in Human Frontal Cortex." *Science* 279 (5355): 1347–51.

Covre, P., A. D. Baddeley, G. J. Hitch, and O. F. A. Bueno. 2019. "Maintaining Task Set Against Distraction: The Role of Working Memory in Multitasking." *Psychology & Neuroscience* 12 (1): 38–52.

Cowan, Nelson. 2010. "The Magical Mystery Four: How Is Working Memory Capacity Limited, and Why?" *Current Directions in Psychological Science* 19 (1): 51–57.

Craik, Fergus I. M. 1994. "Memory Changes in Normal Aging." *Current Directions in Psychological Science* 3 (5): 155–58.

Craik, Fergus I. M., and Cheryl L. Grady. 2002. "Aging, Memory, and Frontal Lobe Functioning." In *Principles of Frontal Lobe Function*, edited by D. T. Stuss and R. T. Knight, 528–40. Oxford: Oxford University Press.

Crews, Frederick. 1995. *The Memory Wars: Freud's Legacy in Dispute.* London: Granta.

Cuc, A., Y. Ozuru, D. Manier, and W. Hirst. 2006. "On the Formation of Collective Memories: The Role of a Dominant Narrator." *Memory & Cognition* 34:752–62.

Curran, T., D. L. Schacter, M. K. Johnson, and R. Spinks. 2001. "Brain Potentials Reflect Behavioral Differences in True and False Recognition." *Journal of Cognitive Neuroscience* 13 (2): 201–16.

Davachi, Lila 2006. "Item, Context and Relational Episodic Encoding in Humans." *Current Opinion in Neurobiology* 16 (6): 693–700.

Davachi, Lila, Jason P. Mitchell, and Anthony D. Wagner. 2003. "Multiple Routes to Memory: Distinct Medial Temporal Lobe Processes Build Item and Source Memories." *Proceedings of the National Academy of Sciences* 100 (4): 2157–62.

Davis, Tyler, Bradley C. Love, and Alison R. Preston. 2012. "Learning the Exception to the Rule: Model-Based fMRI Reveals Specialized Representations for Surprising Category Members." *Cerebral Cortex* 22, no. 2: 260–73.

De Brigard, Felipe, and Natasha Parikh. 2019. "Episodic Counterfactual Thinking." *Current Directions in Psychological Science* 28 (1): 59–66.

Deese, James. 1959. "On the Prediction of Occurrence of Particular Verbal Intrusions in Immediate Recall." *Journal of Experimental Psychology* 58 (1): 17–22. https://doi.org/10.1037/h0046671.

Della Rocchetta, A. I., and B. Milner. 1993. "Strategic Search and Retrieval Inhibition: The Role of the Frontal Lobes." *Neuropsychologia* 31 (6): 503–24.

Den Ouden, H. E., P. Kok, and F. P. De Lange. 2012. "How Prediction Errors Shape Perception, Attention, and Motivation." *Frontiers in Psychology* 3:548.

D'Esposito, M., J. W. Cooney, A. Gazzaley, S. E. Gibbs, and B. R. Postle. 2006. "Is the Prefrontal Cortex Necessary for Delay Task Performance? Evidence from Lesion and fMRI Data." *Journal of the International Neuropsychological Society* 12 (2): 248–60.

D'Esposito, Mark, Bradley R. Postle. 2015. "The Cognitive Neuroscience of Working Memory." *Annual Review of Psychology* 66:115–42.

D'Esposito, Mark, Bradley R. Postle, and Bart Rypma. 2000. "Prefrontal Cortical Contributions to Working Memory: Evidence from Event-Related fMRI Studies." *Experimental Brain Research* 133 (1) (July): 3–11.

Deuker, L., J. L. Bellmund, T. Navarro Schröder, and C. F. Doeller. 2016. "An Event Map of Memory Space in the Hippocampus." *eLife* 5:e16534.

Diamond, Adele. 2006. "The Early Development of Executive Functions." In *Lifespan Cognition: Mechanisms of Change*, edited by E. Bialystok and F. I. M. Craik, 70–95. Oxford: Oxford University Press.

Diana, Rachel A., Andrew P. Yonelinas, and Charan Ranganath. 2007. "Imaging Recollection and Familiarity in the Medial Temporal Lobe: A Three-Component Model." *Trends in Cognitive Sciences* 11 (9): 379–86.

Dickerson, Sally S., and Margaret E. Kemeny. 2004. "Acute Stressors and Cortisol Responses: A Theoretical Integration and Synthesis of Laboratory Research." *Psychological Bulletin* 130 (3): 355.

Diekelmann, S., and J. Born. 2010. "The Memory Function of Sleep." *Nature Reviews Neuroscience* 11 (2): 114–26.

Dimsdale-Zucker, Halle R., Maria E. Montchal, Zachariah M. Reagh, Shao-Fang Wang, Laura A. Libby, and Charan Ranganath. 2022. "Representations of Complex Contexts: A Role for Hippocampus." *Journal of Cognitive Neuroscience* 35 (1): 90–110.

Dimsdale-Zucker, Halle R., and Charan Ranganath. 2018. "Representational Similarity Analyses: A Practical Guide for Functional MRI Applications." *Handbook of Behavioral Neuroscience* 28:509–25.

Dimsdale-Zucker, Halle R., Maureen Ritchey, Arne D. Ekstrom, Andrew P. Yonelinas, and Charan Ranganath. 2018. "CA1 and CA3 Differentially Support Spontaneous Retrieval of Episodic Contexts Within Human Hippocampal Subfields." *Nature Communications* 9 (1): 1–8.

Dixon, R. A. 2011. "Evaluating Everyday Competence in Older Adult Couples: Epidemiological Considerations." *Gerontology* 57 (2): 173–79.

Dobbins, I. G., H. Foley, D. L. Schacter, and A. D. Wagner. 2002. "Executive Control During Episodic Retrieval: Multiple Prefrontal Processes Subserve Source Memory." *Neuron* 35 (5): 989–96.

Douaud, Gwenaëlle, Soojin Lee, Fidel Alfaro-Almagro, Christoph Arthofer, Chaoyue Wang, Paul McCarthy, Frederik Lange, et al. 2022. "SARS-CoV-2 Is Associated with Changes in Brain Structure in UK Biobank." *Nature* 604 (7907): 697–707.

Druzgal, T. J., and M. D'Esposito. 2001. "Activity in Fusiform Face Area Modulated as a Function of Working Memory Load." *Cognitive Brain Research* 10 (3): 355–64.

———. 2003. "Dissecting Contributions of Prefrontal Cortex and Fusiform Face Area to Face Working Memory." *Journal of Cognitive Neuroscience* 15 (6): 771–84.

Duff, M. C., J. Hengst, D. Tranel, and N. J. Cohen. 2006. "Development of Shared Information in Communication Despite Hippocampal Amnesia." *Nature Neuroscience* 9 (1) (January): 140–46. https://doi.org/10.1038/nn1601.

Duff, Melissa C., Jake Kurczek, Rachael Rubin, Neal J. Cohen, and Daniel Tranel. 2013. "Hippocampal Amnesia Disrupts Creative Thinking." *Hippocampus* 23 (12): 1143–49.

Düzel, Emrah, Nico Bunzeck, Marc Guitart-Masip, and Sandra Düzel. 2010. "Novelty-Related Motivation of Anticipation and Exploration by Dopamine (NOMAD): Implications for Healthy Aging." *Neuroscience & Biobehavioral Reviews* 34 (5): 660–69.

Düzel, Emrah, Gabriel Ziegler, David Berron, Anne Maass, Hartmut Schütze, Arturo Cardenas-Blanco, Wenzel Glanz, et al. 2022. "Amyloid Pathology but Not APOE ε4 Status Is Permissive for Tau-Related Hippocampal Dysfunction." *Brain* 145 (4): 1473–85.

Eacott, M. J., and E. A. Gaffan. 2005. "The Roles of Perirhinal Cortex, Postrhinal Cortex, and the Fornix in Memory for Objects, Contexts, and Events in the Rat." *Quarterly Journal of Experimental Psychology Section B* 58 (3–4b): 202–17.

Eagleman, David. 2020. *Livewired: The Inside Story of the Ever-Changing Brain.* Edinburgh: Canongate Books.

Ebbinghaus, Hermann. 1885. *Über das Gedächtnis: Untersuchungen zur experimentellen Psychologie.* Berlin: Duncker & Humblot.

———. 1964. *Memory: A Contribution to Experimental Psychology.* Translated by H. A. Ruger and C. E. Bussenius. New York: Dover. Original work published 1885.

Echterhoff, G., E. T. Higgins, R. Kopietz, and S. Groll. 2008. "How Communication Goals Determine When Audience Tuning Biases Memory." *Journal of Experimental Psychology: General* 137 (1): 3.

Eggins, S., and D. Slade. 2004. *Analysing Casual Conversation.* Sheffield: Equinox Publishing.

Eich, Eric. 1995. "Searching for Mood Dependent Memory." *Psychological Science* 6 (2): 67–75.

Eichenbaum, Howard. 2017. "On the Integration of Space, Time, and Memory." *Neuron* 95 (5): 1007–18.

Eichenbaum, Howard, Andrew Yonelinas, and Charan Ranganath. 2007. "The Medial Temporal Lobe and Recognition Memory." *Annual Review of Neuroscience* 30:123–52.

Ekstrom, Arne D., and Charan Ranganath. 2018. "Space, Time, and Episodic Memory: The Hippocampus Is All Over the Cognitive Map." *Hippocampus* 28 (9): 680–87.

Ericsson, K. Anders, and Walter Kintsch. 1995. "Long-Term Working Memory." *Psychological Review* 102 (2): 211.

Estes, William K. 1955. "Statistical Theory of Spontaneous Recovery and Regression." *Psychological Review* 62:145–54.

Ezzyat, Youssef, and Lila Davachi. 2011. "What Constitutes an Episode in Episodic Memory?" *Psychological Science* 22 (2): 243–52.

Fandakova, Yana, and Matthias J. Gruber. 2021. "States of Curiosity and Interest Enhance Memory Differently in Adolescents and in Children." *Developmental Science* 24 (1): e13005.

Feduccia, Allison A., and Michael C. Mithoefer. 2018. "MDMA-Assisted Psychotherapy for PTSD: Are Memory Reconsolidation and Fear Extinction Underlying Mechanisms?" *Progress in Neuro-psychopharmacology and Biological Psychiatry* 84:221–28.

Fillit, Howard M., Robert N. Butler, Alan W. O'Connell, Marilyn S. Albert, James E. Birren, Carl W. Cotman, William T. Greenough, et al. 2002. "Achieving and Maintaining Cognitive Vitality with Aging." *Mayo Clinic Proceedings* 77 (7).

Fivush, Robyn. 2004. "Voice and Silence: A Feminist Model of Autobiographical Memory." In *The Development of the Mediated Mind: Sociocultural Context and Cognitive Development*, edited by Joan M. Lucariello, Judith A. Hudson, Robyn Fivush, and Patricia J. Bauer, 79–100. Mahwah, NJ: Lawrence Erlbaum.

Fivush, Robyn. 2008. "Remembering and Reminiscing: How Individual Lives Are Constructed in Family Narratives." *Memory Studies* 1 (1): 49–58.

Foerster, Otfrid, and Wilder Penfield. 1930. "The Structural Basis of Traumatic Epilepsy and Results of Radical Operation." *Brain* 53:99–119.

Franklin, Nicholas T., Kenneth A. Norman, Charan Ranganath, Jeffrey M. Zacks, and Samuel J. Gershman. 2020. "Structured Event Memory: A Neuro-symbolic Model of Event Cognition." *Psychological Review* 127 (3): 327.

Frenda, Steven J., Eric D. Knowles, William Saletan, and Elizabeth F. Loftus. 2013. "False Memories of Fabricated Political Events." *Journal of Experimental Social Psychology* 49 (2): 280–86.

Freyd, Jennifer J. 1998. "Science in the Memory Debate." *Ethics & Behavior* 8 (2): 101–13.

Fuster, Joaquin M. 1980. *The Prefrontal Cortex: Anatomy, Physiology, and Neuropsychology of the Frontal Lobe.* New York: Raven Press.

Gaesser, Brendan, and Daniel L. Schacter. 2014. "Episodic Simulation and Episodic Memory Can Increase Intentions to Help Others." *Proceedings of the National Academy of Sciences* 111 (12): 4415–20.

Galli, G., M. Sirota, M. J. Gruber, B. E. Ivanof, J. Ganesh, M. Materassi, A. Thorpe, V. Loaiza, M. Cappelletti, and F. I. Craik. 2018. "Learning Facts During Aging: The Benefits of Curiosity." *Experimental Aging Research* 44 (4): 311–28.

Gazzaley, A., and L. D. Rosen. 2016. *The Distracted Mind: Ancient Brains in a High-Tech World.* Cambridge, MA: MIT Press.

Gershberg, Felicia B., and Arthur P. Shimamura. 1995. "Impaired Use of Organizational Strategies in Free Recall Following Frontal Lobe Damage." *Neuropsychologia* 33 (10): 1305–33.

Gershman, S. J., P. E. Balbi, C. Balbi, C. R. Gallistel, and J. Gunawardena. 2021. "Reconsidering the Evidence for Learning in Single Cells." *eLife* 10:e61907. https://doi.org/10.7554/eLife.61907.

Gershman, S. J., M. H. Monfils, K. A. Norman, and Y. Niv. 2017. "The Computational Nature of Memory Modification." *eLife* 6:e23763.

Gerstorf, D., C. A. Hoppman, K. J. Anstey, and M. A. Luszcz. 2009. "Dynamic Links of Cognitive Functioning Among Married Couples: Longitudinal Evidence from the Australian Longitudinal Study of Ageing." *Psychology and Aging* 24 (2): 296.

Geva-Sagiv, Maya, and Yuval Nir. 2019. "Local Sleep Oscillations: Implications for Memory Consolidation." *Frontiers in Neuroscience* 13:813.

Ghansah, Rachel Kaadzi. 2011. "The B-Boy's Guide to the Galaxy." *Transition: An International Review* 104:122–36.

Ghetti, Simona. 2017. "Development of Item-Space and Item-Time Binding." *Current Opinion in Behavioral Sciences* 17:211–16.

Ghetti, Simona, and Paola Castelli. 2006. "Developmental Differences in False-Event Rejection: Effects of Memorability-Based Warnings." *Memory* 14 (6): 762–76.

Ghetti, Simona, Robin S. Edelstein, Gail S. Goodman, Ingrid M. Cordòn, Jodi A. Quas, Kristen Weede Alexander, Allison D. Redlich, and David P. H. Jones. 2006. "What Can Subjective Forgetting Tell Us About Memory for Childhood Trauma?" *Memory & Cognition* (34): 1011–25.

Gobet, Fernand, and Herbert A. Simon. 1998. "Expert Chess Memory: Revisiting the Chunking Hypothesis." *Memory* 6 (3): 225–55.

Godoy, Lívea Dornela, Matheus Teixeira Rossignoli, Polianna Delfino-Pereira, Norberto Garcia-Cairasco, and Eduardo Henrique de Lima Umeoka. 2018. "A Comprehensive Overview on Stress Neurobiology: Basic Concepts and Clinical Implications." *Frontiers in Behavioral Neuroscience* 12:127.

Goldman-Rakic, Patricia S. 1984. "The Frontal Lobes: Uncharted Provinces of the Brain." *Trends in Neurosciences* 7 (11): 425–29.

———. 1987. "Circuitry of Primate Prefrontal Cortex and Regulation of Behavior by Representational Memory." In *Higher Functions of the Brain: The Nervous System; Handbook of Physiology*, edited by F. Plum, 5:373–417, section 1. Bethesda, MD: American Physiological Society.

Gonsalves, Brian, and Ken A. Paller. 2000. "Neural Events That Underlie Remembering Something That Never Happened." *Nature Neuroscience* 3 (12): 1316–21.

Gonsalves, Brian, Paul J. Reber, Darren R. Gitelman, Todd B. Parrish, M.-Marsel Mesulam, and Ken A. Paller. 2004. "Neural Evidence That Vivid Imagining Can Lead to False Remembering." *Psychological Science* 15 (10): 655–60.

Gooding, R. 2004. "The Trashing of John McCain." *Vanity Fair*, November 2004. https://www.vanityfair.com/news/2004/11/mccain200411.

Goodman, Gail S., Simona Ghetti, Jodi A. Quas, Robin S. Edelstein, Kristen Weede Alexander, Allison D. Redlich, Ingrid M. Cordon, and David P. H. Jones. 2003. "A Prospective Study of Memory for Child Sexual Abuse: New Findings Relevant to the Repressed-Memory Controversy." *Psychological Science* 14 (2): 113–18.

Gopnik, Alison. 2020. "Childhood as a Solution to Explore-Exploit Tensions." *Philosophical Transactions of the Royal Society B* 375 (1803): 20190502.

Gotlib, I. H., C. Ranganath, and J. P. Rosenfeld. 1998. "Frontal EEG Alpha Asymmetry, Depression, and Cognitive Functioning." *Cognition and Emotion* 12 (3): 449–78. https://doi.org/10.1080/026999398379673.

Grady, Cheryl. 2012. "The Cognitive Neuroscience of Ageing." *Nature Reviews Neuroscience* 13 (7): 491–505.

Gray, J. A. 1982. "Précis of the Neuropsychology of Anxiety: An Enquiry into the Functions of the Septo-Hippocampal System." *Behavioral and Brain Sciences* 5, no. 3: 469–84.

Greeley, G. D., V. Chan, H.-Y. Choi, and S. Rajaram. 2023. "Collaborative Recall and the Construction of Collective Memory Organization: The Impact of Group Structure." *Topics in Cognitive Science.* https://doi.org/10.1111/tops.12639.

Greeley, G. D., and S. Rajaram. 2023. "Collective Memory: Collaborative Recall Synchronizes What and How People Remember." *WIREs Cognitive Science.* https://doi.org/10.1002/wcs.1641.

Griego, A. W., J. N. Datzman, S. M. Estrada, and S. S. Middlebrook. 2019. "Suggestibility and False Memories in Relation to Intellectual Disability and Autism Spectrum Disorder: A Meta-analytic Review." *Journal of Intellectual Disability Research* 63 (12): 1464–74.

Grill-Spector, K., R. Henson, and A. Martin. 2006. "Repetition and the Brain: Neural Models of Stimulus-Specific Effects." *Trends in Cognitive Sciences* 10 (1): 14–23.

Gross, J. J., and L. Feldman Barrett. 2011. "Emotion Generation and Emotion Regulation: One or Two Depends on Your Point of View." *Emotion Review* 3 (1): 8–16.

Grossberg, Stephen. 1976. "Adaptive Pattern Classification and Universal Recoding. II: Feedback, Expectation, Olfaction, and Illusions." *Biological Cybernetics* 23:187–202.

Gruber, Matthias J., Bernard D. Gelman, and Charan Ranganath. 2014. "States of Curiosity Modulate Hippocampus-Dependent Learning via the Dopaminergic Circuit." *Neuron* 84 (2): 486–96.

Gruber, Matthias J., and Charan Ranganath. 2019. "How Curiosity Enhances Hippocampus-Dependent Memory: The Prediction, Appraisal, Curiosity, and Exploration (PACE) Framework." *Trends in Cognitive Sciences* 23 (12): 1014–25.

Gruber, M. J., M. Ritchey, S. F. Wang, M. K. Doss, and C. Ranganath. 2016. "Post-Learning Hippocampal Dynamics Promote Preferential Retention of Rewarding Events." *Neuron* 89 (5): 1110–20.

Grunwald, Thomas, Heinz Beck, Klaus Lehnertz, Ingmar Blümcke, Nico Pezer, Marta Kutas, Martin Kurthen, et al. 1999. "Limbic P300s in Temporal Lobe Epilepsy with and Without Ammon's Horn Sclerosis." *European Journal of Neuroscience* 11(6): 1899–906.

Grunwald, T., C. E. Elger, K. Lehnertz, D. Van Roost, and H. J. Heinze. 1995. "Alterations of Intrahippocampal Cognitive Potentials in Temporal Lobe Epilepsy." *Electroencephalography and Clinical Neurophysiology* 95 (1): 53–62.

Haber, S. N. 2011. "Neuroanatomy of Reward: A View from the Ventral Striatum." Chap. 11 in *Neurobiology of Sensation and Reward*, edited by J. A. Gottfried, 235. Boca Raton, FL: CRC Press / Taylor & Francis.

Hagwood, Scott. 2006. *Memory Power: You Can Develop a Great Memory—America's Grand Master Shows You How*. New York: Simon & Schuster.

Halbwachs, Maurice. 1992. *On Collective Memory*. Translated by Lewis A. Coser. Chicago: University of Chicago Press.

Hannula, D. E., R. R. Althoff, D. E. Warren, L. Riggs, N. J .Cohen, and J. D. Ryan. 2010. "Worth a Glance: Using Eye Movements to Investigate the Cognitive Neuroscience of Memory." *Frontiers in Human Neuroscience* 4:166.

Hannula, D. E., L. A. Libby, A. P. Yonelinas, and C. Ranganath. 2013. "Medial Temporal Lobe Contributions to Cued Retrieval of Items and Contexts." *Neuropsychologia* 51 (12): 2322–32.

Hannula, Deborah E., and Charan Ranganath. 2009. "The Eyes Have It: Hippocampal Activity Predicts Expression of Memory in Eye Movements." *Neuron* 63 (5): 592–99.

Hannula, Deborah E., Jennifer D. Ryan, Daniel Tranel, and Neal J. Cohen. 2007. "Rapid Onset Relational Memory Effects Are Evident in Eye Movement Behavior, but Not in Hippocampal Amnesia." *Journal of Cognitive Neuroscience* 19 (10): 1690–705.

Harris, C. B., A. J. Barnier, J. Sutton, and P. G. Keil. 2014. "Couples as Socially Distributed Cognitive Systems: Remembering in Everyday Social and Material Contexts." *Memory Studies* 7 (3): 285–97.

Harris, C. B., P. G. Keil, J. Sutton, A. J. Barnier, and D. J. McIlwain. 2011. "We Remember, We Forget: Collaborative Remembering in Older Couples." *Discourse Processes* 48 (4): 267–303.

Harrison, George. 2007. *I Me Mine*. San Francisco: Chronicle Books.

Harrison, Neil A., Kate Johnston, Federica Corno, Sarah J. Casey, Kimberley Friedner, Kate Humphreys, Eli J. Jaldow, Mervi Pitkanen, and Michael D. Kopelman. 2017. "Psychogenic Amnesia: Syndromes, Outcome, and Patterns of Retrograde Amnesia." *Brain* 140 (9): 2498–510.

Hart, C. L. 2017. "Viewing Addiction as a Brain Disease Promotes Social Injustice." *Nature Human Behaviour* 1 (3): 55.

Hasher, Lynn, David Goldstein, and Thomas Toppino. 1977. "Frequency and the Conference of Referential Validity." *Journal of Verbal Learning and Verbal Behavior* 16 (1): 107–12.

Hashtroudi, Shahin, Marcia K. Johnson, and Linda D. Chrosniak. 1989. "Aging and Source Monitoring." *Psychology and Aging* 4 (1): 106.

Hassabis, D., D. Kumaran, and E. A. Maguire. 2007. "Using Imagination to Understand the Neural Basis of Episodic Memory." *Journal of Neuroscience* 27 (52): 14365–74.

Hassabis, Demis, Dharshan Kumaran, Seralynne D. Vann, and Eleanor A. Maguire. 2007. "Patients with Hippocampal Amnesia Cannot Imagine New Experiences." *Proceedings of the National Academy of Sciences* 104 (5): 1726–31.

Hauer, B. J., and I. Wessel. 2006. "Retrieval-Induced Forgetting of Autobiographical Memory Details." *Cognition and Emotion* 20 (3–4): 430–47.

Haxby, J. V., M. I. Gobbini, M. L. Furey, A. Ishai, J. L. Schouten, and P. Pietrini. 2001. "Distributed and Overlapping Representations of Faces and Objects in Ventral Temporal Cortex." *Science* 293 (5539): 2425–30.

Hayes, Taylor R., and John M. Henderson. 2021. "Looking for Semantic Similarity: What a Vector-Space Model of Semantics Can Tell Us About Attention in Real-World Scenes." *Psychological Science* 32 (8): 1262–70.

Healey, M. Karl, Nicole M. Long, and Michael J. Kahana. 2019. "Contiguity in Episodic Memory." *Psychonomic Bulletin & Review* 26 (3): 699–720.

Hebb, Donald O. 1949. *The Organization of Behavior: A Neuropsychological Theory.* Hoboken, NJ: Wiley.

Heilig, M., J. MacKillop, D. Martinez, J. Rehm, L. Leggio, and L. J. Vanderschuren. 2021. "Addiction as a Brain Disease Revised: Why It Still Matters, and the Need for Consilience." *Neuropsychopharmacology* 46 (10): 1715–23.

Henderson, John M., and Taylor R. Hayes. 2017. "Meaning-Based Guidance of Attention in Scenes as Revealed by Meaning Maps." *Nature Human Behaviour* 1 (10): 743–47.

Henderson, John M., Taylor R. Hayes, Candace E. Peacock, and Gwendolyn Rehrig. 2019. "Meaning and Attentional Guidance in Scenes: A Review of the Meaning Map Approach." *Vision* 3 (2): 19.

Hendrickson, Carolyn W., Reeva J. Kimble, and Daniel P. Kimble. 1969. "Hippocampal Lesions and the Orienting Response." *Journal of Comparative and Physiological Psychology* 67 (2, pt. 1): 220.

Henkel, Linda A. 2014. "Point-and-Shoot Memories: The Influence of Taking Photos on Memory for a Museum Tour." *Psychological Science* 25 (2): 396–402.

Henkel, Linda A., and Kimberly J. Coffman. 2004. "Memory Distortions in Coerced False Confessions: A Source Monitoring Framework Analysis." *Applied Cognitive Psychology: The Official Journal of the Society for Applied Research in Memory and Cognition* 18 (5): 567–88.

Herculano-Houzel, Suzana. 2012. "The Remarkable, Yet Not Extraordinary, Human Brain as a Scaled-Up Primate Brain and Its Associated Cost." *Proceedings of the National Academy of Sciences* 109 (suppl. 1): 10661–68.

Higgins, E. T., and W. S. Rholes. 1978. " 'Saying Is Believing': Effects of Message Modification on Memory and Liking for the Person Described." *Journal of Experimental Social Psychology* 14 (4): 363–78.

Hilgetag, C. C., M. A. O'Neill, and M. P. Young. 1997. "Optimization Analysis of Complex Neuroanatomical Data." In *Computational Neuroscience*, edited by J. M. Bower. Boston: Springer.

Hirst, William. 2010. "The Contribution of Malleability to Collective Memory." In *The Cognitive Neuroscience of Mind: A Tribute to Michael S. Gazzaniga*, edited by Patricia A. Reuter-Lorenz, Kathleen Baynes, George R. Mangun, and Elizabeth A. Phelps, 139–53. Cambridge, MA: MIT Press.

Hirst, William, and Bruce T. Volpe. 1988. "Memory Strategies with Brain Damage." *Brain and Cognition* 8 (3): 379–408.

Ho, J. W., D. L. Poeta, T. K. Jacobson, T. A. Zolnik, G. T, Neske, B. W. Connors, and R. D. Burwell. 2015. "Bidirectional Modulation of Recognition Memory." *Journal of Neuroscience* 35 (39) (September 30): 13323–35.

Horton, W. S., and R. J. Gerrig. 2005. "Conversational Common Ground and Memory Processes in Language Production." *Discourse Processes* 40:1–35.

———. 2016. "Revisiting the Memory-Based Processing Approach to Common Ground." *Topics in Cognitive Science* 8:780–95.

Howe, Mark L., and Mary L. Courage. 1993. "On Resolving the Enigma of Infantile Amnesia." *Psychological Bulletin* 113 (2): 305.

Hsieh, Liang-Tien, Matthias J. Gruber, Lucas J. Jenkins, and Charan Ranganath. 2014. "Hippocampal Activity Patterns Carry Information About Objects in Temporal Context." *Neuron* 81 (5): 1165–78.

Hu, X., J. W. Antony, J. D. Creery, I. M. Vargas, G. V. Bodenhausen, and K. A. Paller. 2015. "Unlearning Implicit Social Biases During Sleep." *Science* 348 (6238): 1013–15.

Hu, X., L. Y. Cheng, M. H. Chiu, and K. A. Paller. 2020. "Promoting Memory Consolidation During Sleep: A Meta-analysis of Targeted Memory Reactivation." *Psychological Bulletin* 146 (3): 218.

Huff, Markus, Frank Papenmeier, Annika E. Maurer, Tino G. K. Meitz, Bärbel Garsoffky, and Stephan Schwan. 2017. "Fandom Biases Retrospective Judgments Not Perception." *Scientific Reports* 7 (1): 1–8.

Hughlings-Jackson, John. 1888. "On a Particular Variety of Epilepsy ('Intellectual Aura'), One Case with Symptoms of Organic Brain Disease." *Brain* 11 (2): 179–207.

Hupbach, Almut, Oliver Hardt, Rebecca Gomez, and Lynn Nadel. 2008. "The Dynamics of Memory: Context-Dependent Updating." *Learning & Memory* 15, no. 8: 574–79.

Inbau, F. E., J. E. Reid, J. P. Buckley, and B. C. Jayne. 2001. *Criminal Interrogation and Confessions*. 4th ed. Gaithersburg, MD: Aspen.

Jacobsen, C. F. 1936. "Studies of Cerebral Function in Primates. I. The Functions of the Frontal Association Areas in Monkeys." In *Comparative Psychology Monographs*, vol. 13. Baltimore: Williams & Wilkins.

Janata, Petr. 2009. "The Neural Architecture of Music-Evoked Autobiographical Memories." *Cerebral Cortex* 19 (11): 2579–94.

Jansari, Ashok, and Alan J. Parkin. 1996. "Things That Go Bump in Your Life: Explaining the Reminiscence Bump in Autobiographical Memory." *Psychology and Aging* 11 (1): 85.

Janssen, Steve M. J., Jaap M. J. Murre, and Martijn Meeter. 2008. "Reminiscence Bump in Memory for Public Events." *European Journal of Cognitive Psychology* 20 (4): 738–64.

Jardine, K. H., A. E. Huff, C. E. Wideman, S. D. McGraw, and B. D. Winter. 2022. "The Evidence for and Against Reactivation-Induced Memory Updating in Humans and Nonhuman Animals." *Neuroscience & Biobehavioral Reviews* 136:104598.

Jenkins, Lucas J., and Charan Ranganath. 2010. "Prefrontal and Medial Temporal Lobe Activity at Encoding Predicts Temporal Context Memory." *Journal of Neuroscience* 30 (46): 15558–65.

Jha, A. 2021. *Peak Mind: Find Your Focus, Own Your Attention, Invest 12 Minutes a Day.* London: Hachette UK.

Jia, Jianping, Tan Zhao, Zhaojun Liu, Yumei Liang, Fangyu Li, Yan Li, Wenying Liu, et al. 2023. "Association Between Healthy Lifestyle and Memory Decline in Older Adults: 10 Year, Population Based, Prospective Cohort Study." *BMJ* 380:e072691.

Johnson, Elliott Gray, Lindsey Mooney, Katharine Graf Estes, Christine Wu Nordahl, and Simona Ghetti. 2021. "Activation for Newly Learned Words in Left Medial-Temporal Lobe during Toddlers' Sleep Is Associated with Memory for Words." *Current Biology* 31, no. 24: 5429–38.

Johnson, M. H. 2001. "Functional Brain Development in Humans." *Nature Reviews Neuroscience* 2 (7): 475–83.

Johnson, M. K., M. A. Foley, A. G. Suengas, and C. L. Raye. 1988. "Phenomenal Characteristics of Memories for Perceived and Imagined Autobiographical Events." *Journal of Experimental Psychology: General* 117 (4): 371.

Johnson, Marcia K., Shahin Hashtroudi, and D. Stephen Lindsay. 1993. "Source Monitoring." *Psychological Bulletin* 114 (1): 3–28.

Johnson, M. K., S. M. Hayes, M. D'Esposito, and C. L. Raye. 2000. "Confabulation." In *Handbook of Neuropsychology: Memory and Its Disorders*, edited by L. S. Cermak, 383–407. Amsterdam: Elsevier.

Johnson, M. K., J. Kounios, and S. F. Nolde. 1997. "Electrophysiological Brain Activity and Memory Source Monitoring." *NeuroReport* 8 (5): 1317–20.

Johnson, M. K., S. F. Nolde, M. Mather, J. Kounios, D. L. Schacter, and T. Curran. 1997. "The Similarity of Brain Activity Associated with True and False Recognition Memory Depends on Test Format." *Psychological Science* 8 (3): 250–57.

Johnson, Marcia K., and Carol L. Raye. 1981. "Reality Monitoring." *Psychological Review* 88 (1): 67–85.

Johnson, Reed. 2017. "The Mystery of S., the Man with an Impossible Memory." *New Yorker,* August 12, 2017. https://www.newyorker.com/books/page-turner/the-mystery-of-s-the-man-with-an-impossible-memory.

Jonides, J., R. L. Lewis, D. E. Nee, C. A. Lustig, M. G. Berman, and K. S. Moore.

2008. "The Mind and Brain of Short-Term Memory." *Annual Review of Psychology* 59:193–224.

Jonker, T. R., H. Dimsdale-Zucker, M. Ritchey, A. Clarke, and C. Ranganath. 2018. "Neural Reactivation in Parietal Cortex Enhances Memory for Episodically Linked Information." *Proceedings of the National Academy of Sciences* 115 (43): 11084–89.

Jonker, T. R., P. Seli, and C. M. MacLeod. 2013. "Putting Retrieval-Induced Forgetting in Context: An Inhibition-Free, Context-Based Account." *Psychological Review* 120 (4): 852.

Joo, H. R., and L. M. Frank. 2018. "The Hippocampal Sharp-Wave Ripple in Memory Retrieval for Immediate Use and Consolidation." *Nature Reviews Neuroscience* 19 (12): 744–57.

Josselyn, Sheena A., Stefan Köhler, and Paul W. Frankland. 2017. "Heroes of the Engram." *Journal of Neuroscience* 37 (18): 4647–57.

Kafkas, A., and D. Montaldi. 2018. "How Do Memory Systems Detect and Respond to Novelty?" *Neuroscience Letters* 680:60–68.

Kahneman, Daniel. 2011. *Thinking, Fast and Slow.* New York: Macmillan.

Kahneman, Daniel, and Jason Riis. 2005. "Living, and Thinking About It: Two Perspectives on Life." Chap. 11 in *The Science of Well-Being*, edited by Felicia A. Huppert, Nick Baylis, and Barry Keverne, 285–304. Oxford: Oxford University Press.

Kandel, Eric R., Yadin Dudai, and Mark Mayford. 2014. "The Molecular and Systems Biology of Memory." *Cell* 157 (1): 163–86. https://doi.org/10.1016/j.cell.2014.03.001.

Kang, M. J., M. Hsu, I. M. Krajbich, G. Loewenstein, S. M. McClure, J. T. Y. Wang, and C. F. Camerer. 2009. "The Wick in the Candle of Learning: Epistemic Curiosity Activates Reward Circuitry and Enhances Memory." *Psychological Science* 20 (8): 963–73.

Kant, Immanuel. 1899. *Critique of Pure Reason.* Translated by J. M. D. Meiklejohn. Willey Book Company. https://doi.org/10.1037/11654-000.

Kaplan, R., M. H. Adhikari, R. Hindriks, D. Mantini, Y. Murayama, N. K. Logothetis, and G. Deco. 2016. "Hippocampal Sharp-Wave Ripples Influence Selective Activation of the Default Mode Network." *Current Biology* 26 (5): 686–91.

Karpicke, Jeffrey D., and Henry L. Roediger III. 2008. "The Critical Importance of Retrieval for Learning." *Science* 319 (5865): 966–68.

Kashima, Yoshihisa. 2000. "Maintaining Cultural Stereotypes in the Serial Reproduction of Narratives." *Personality and Social Psychology Bulletin* 26 (5): 594–604.

Kassin, Saul M. 2008. "False Confessions: Causes, Consequences, and Implications for Reform." *Current Directions in Psychological Science* 17 (4): 249–53.

Kelley, C. M., and L. L. Jacoby. 1990. "The Construction of Subjective Experience: Memory Attributions." *Mind & Language* 5 (1): 49–68.

Ketz, N., S. G. Morkonda, and R .C. O'Reilly. 2013. "Theta Coordinated Error-Driven Learning in the Hippocampus." *PloS Computational Biology* 9 (6): e1003067.

Kidd, C., and B. Y. Hayden. 2015. "The Psychology and Neuroscience of Curiosity." *Neuron* 88 (3): 449–60.

Kirschbaum, Clemens, Karl-Martin Pirke, and Dirk H. Hellhammer. 1993. "The 'Trier Social Stress Test'—a Tool for Investigating Psychobiological Stress Responses in a Laboratory Setting." *Neuropsychobiology* 28 (1–2): 76–81.

Kloft, Lilian, Lauren A. Monds, Arjan Blokland, Johannes G. Ramaekers, and Henry Otgaar. 2021. "Hazy Memories in the Courtroom: A Review of Alcohol and Other Drug Effects on False Memory and Suggestibility." *Neuroscience & Biobehavioral Reviews* 124:291–307.

Kloft, Lilian, Henry Otgaar, Arjan Blokland, Lauren A. Monds, Stefan W. Toennes, Elizabeth F. Loftus, and Johannes G. Ramaekers. 2020. "Cannabis Increases Susceptibility to False Memory." *Proceedings of the National Academy of Sciences* 117 (9): 4585–89.

Knierim, J. J., I. Lee, and E. L. Hargreaves. 2006. "Hippocampal Place Cells: Parallel Input Streams, Subregional Processing, and Implications for Episodic Memory." *Hippocampus* 16 (9): 755–64.

Knight, Erik L., and Pranjal H. Mehta. 2017. "Hierarchy Stability Moderates the Effect of Status on Stress and Performance in Humans." *Proceedings of the National Academy of Sciences* 114 (1): 78–83.

Knight, Robert T. 1984. "Decreased Response to Novel Stimuli After Prefrontal Lesions in Man." *Electroencephalography and Clinical Neurophysiology* 59 (1): 9–20. doi:10.1016/0168-5597(84)90016-9.

———. 1996. "Contribution of Human Hippocampal Region to Novelty Detection." *Nature* 383 (6597): 256–59.

Knutson, B., C. M. Adams, G. W. Fong, and D. Hommer. 2001. "Anticipation of Increasing Monetary Reward Selectively Recruits Nucleus Accumbens." *Journal of Neuroscience* 21 (16) (August 15): RC159. https://doi.org/10.1523/JNEUROSCI.21-16-j0002.2001.

Koppel, Jonathan, Dana Wohl, Robert Meksin, and William Hirst. 2014. "The Effect of Listening to Others Remember on Subsequent Memory: The Roles of Expertise and Trust in Socially Shared Retrieval-Induced Forgetting and Social Contagion." *Social Cognition* 32 (2): 148–80.

Krause, A. J., E. B. Simon, B. A. Mander, S. M. Greer, J. M. Saletin, A. N. Goldstein-Piekarski, and M. P. Walker. 2017. "The Sleep-Deprived Human Brain." *Nature Reviews Neuroscience* 18 (7) (July): 404–18.

Kriegeskorte, Nikolaus, Marieke Mur, and Peter A. Bandettini. 2008. "Representational Similarity Analysis—Connecting the Branches of Systems Neuroscience." *Frontiers in Systems Neuroscience* 2:4.

Krumhansl, Carol Lynne, and Justin Adam Zupnick. 2013. "Cascading Reminiscence Bumps in Popular Music." *Psychological Science* 24 (10): 2057–68.

LaBar, Kevin S., and Roberto Cabeza. 2006. "Cognitive Neuroscience of Emotional Memory." *Nature Reviews Neuroscience* 7 (1): 54–64.

LeDoux, Joseph. 2012. "Rethinking the Emotional Brain." *Neuron* 73 (4): 653–76.

Lensvelt-Mulders, G., O. van der Hart, J. M. van Ochten, M. J. van Son, K. Steele, and L. Breeman. 2008. "Relations Among Peritraumatic Dissociation and Posttraumatic Stress: A Meta-analysis." *Clinical Psychology Review* 28 (7): 1138–51.

LePort, Aurora K. R., Aaron T. Mattfeld, Heather Dickinson-Anson, James H. Fallon, Craig E. L. Stark, Frithjof Kruggel, Larry Cahill, and James L. McGaugh. 2012. "Behavioral and Neuroanatomical Investigation of Highly Superior Autobiographical Memory (HSAM)." *Neurobiology of Learning and Memory* 98 (1): 78–92.

LePort, Aurora K. R., S. M. Stark, J. L. McGaugh, and C. E. L. Stark. 2016. "Highly Superior Autobiographical Memory: Quality and Quantity of Retention over Time." *Frontiers in Psychology* 6, article 2017.

Levin, Daniel T. 2000. "Race as a Visual Feature: Using Visual Search and Perceptual Discrimination Tasks to Understand Face Categories and the Cross-Race Recognition Deficit." *Journal of Experimental Psychology: General* 129 (4): 559.

Levitin, Daniel J. 2014. *The Organized Mind: Thinking Straight in the Age of Information Overload.* New York: Penguin.

———. 2020. *Successful Aging: A Neuroscientist Explores the Power and Potential of Our Lives.* New York: Penguin.

Lewis, Penelope A., and Simon J. Durrant. 2011. "Overlapping Memory Replay During Sleep Builds Cognitive Schemata." *Trends in Cognitive Sciences* 15 (8): 343–51. doi:10.1016/j.tics.2011.06.004.

Libby, Laura A., Deborah E. Hannula, and Charan Ranganath. 2014. "Medial Temporal Lobe Coding of Item and Spatial Information During Relational Binding in Working Memory." *Journal of Neuroscience* 34 (43): 14233–42.

Libby, Laura A., Zachariah M. Reagh, Nichole R. Bouffard, J. Daniel Ragland, and Charan Ranganath. 2019. "The Hippocampus Generalizes Across Memories That Share Item and Context Information." *Journal of Cognitive Neuroscience* 31 (1): 24–35.

Liberzon, Israel, and James L. Abelson. 2016. "Context Processing and the Neurobiology of Post-Traumatic Stress Disorder." *Neuron* 92 (1): 14–30.

Lisman, John E., and Anthony A. Grace. 2005. "The Hippocampal-VTA Loop: Controlling the Entry of Information into Long-Term Memory." *Neuron* 46 (5): 703–13.

Liu, Xiaonan L., Randall C. O'Reilly, and Charan Ranganath. 2021. "Effects of Retrieval Practice on Tested and Untested Information: Cortico-hippocampal Interactions and Error-Driven Learning." *Psychology of Learning and Motivation* 75:125–55.

Liu, Xiaonan L., and Charan Ranganath. 2021. "Resurrected Memories: Sleep-Dependent Memory Consolidation Saves Memories from Competition Induced by Retrieval Practice." *Psychonomic Bulletin & Review* 28 (6): 2035–44.

Liu, Xiaonan L., Charan Ranganath, and Randall C. O'Reilly. 2024. "A Complementary Learning Systems Model of How Sleep Moderates Retrieval Practice Effects." *Psychonomic Bulletin & Review,* doi: 10.3758/s13423-024-02489-1.

Lockhart, S. N., A. B. Mayda, A. E. Roach, O. Fletcher, O. Carmichael, P. Maillard, C. G. Schwarz, A. P. Yonelinas, C. Ranganath, and C. DeCarli. 2012. "Episodic Memory Function Is Associated with Multiple Measures of White Matter Integrity in Cognitive Aging." *Frontiers in Human Neuroscience* 6 (March 16): 56.

Loewenstein, George. 1994. "The Psychology of Curiosity: A Review and Reinterpretation." *Psychological Bulletin* 116 (1): 75.

Loewenstein, George, and David Schkade. 1999. "Wouldn't It Be Nice?" In *Predicting Future Feelings. Well-Being: The Foundations of Hedonic Psychology,* edited by D. Kahneman, E. Diener, and N. Schwarz, 85–105. New York: Russell Sage Foundation.

Loftus, Elizabeth F. 2005. "Planting Misinformation in the Human Mind: A 30-Year Investigation of the Malleability of Memory." *Learning & Memory* 12 (4): 361–66.

Loftus, Elizabeth F., and Deborah Davis. 2006. "Recovered Memories." *Annual Review of Clinical Psychology* 2:469–98.

Loftus, Elizabeth F., and John C. Palmer. 1974. "Reconstruction of Automobile Destruction: An Example of the Interaction Between Language and Memory." *Journal of Verbal Learning and Verbal Behavior* 13 (5): 585–89.

Loftus, Elizabeth F., and Jacqueline E. Pickrell. 1995. "The Formation of False Memories." *Psychiatric Annals* 25 (12): 720–25.

Love, B. C., Medin, D. L., and Gureckis, T. M. 2009. "SUSTAIN: A Network Model of Category Learning." *Psychological Review* 111, no. 2: 309–32.

Lu, Q., U. Hasson, and K. A. Norman. 2022. "A Neural Network Model of When to Retrieve and Encode Episodic Memories." *eLife* 11:e74445.

Luck, Steven J., and Edward K. Vogel. 2013. "Visual Working Memory Capacity:

From Psychophysics and Neurobiology to Individual Differences." *Trends in Cognitive Sciences* 17 (8): 391–400.

Luhmann, C. C., and S. Rajaram. 2015. "Memory Transmission in Small Groups and Large Networks: An Agent-Based Model." *Psychological Science* 26 (12): 1909–17.

Lupien, S. J., B. S. McEwen, M. R. Gunnar, and C. Heim. 2009. "Effects of Stress Throughout the Lifespan on the Brain, Behaviour and Cognition." *Nature Reviews Neuroscience* 10 (6): 434–45.

Luria, A. R. 1968. *The Mind of the Mnemonist.* New York: Basic Books.

Lyons, Anthony, and Yoshihisa Kashima. 2001. "The Reproduction of Culture: Communication Processes Tend to Maintain Cultural Stereotypes." *Social Cognition* 19 (3): 372–94.

———. 2003. "How Are Stereotypes Maintained Through Communication? The Influence of Stereotype Sharedness." *Journal of Personality and Social Psychology* 85 (6): 989.

MacLeod, Colin M. 2024. "Interference Theory: History and Current Status." Chap. 40 in *The Oxford Handbook of Human Memory*, vol. 2, edited by M. J. Kahana and A. D. Wagner, 1173–1208. Oxford: Oxford University Press.

MacMillan, Amanda. 2017. "The Downside of Having an Almost Perfect Memory." *Time*, December 8, 2017. https://time.com/5045521/highly-superior-autobiographical-memory-hsam/.

Madore, Kevin P., Preston P. Thakral, Roger E. Beaty, Donna Rose Addis, and Daniel L. Schacter. 2019. "Neural Mechanisms of Episodic Retrieval Support Divergent Creative Thinking." *Cerebral Cortex* 29 (1): 150–66.

Maguire, Eleanor A., Faraneh Vargha-Khadem, and Demis Hassabis. 2010. "Imagining Fictitious and Future Experiences: Evidence from Developmental Amnesia." *Neuropsychologia* 48 (11): 3187–92.

Mandler, George. 1980. "Recognizing: The Judgment of Previous Occurrence." *Psychological Review* 87 (3): 252.

Manning, J. R., K. A. Norman, and M. J. Kahana. 2014. "The Role of Context in Episodic Memory." In *The Cognitive Neurosciences V*, edited by M. Gazzaniga and R. Mangun. Cambridge, MA: MIT Press.

Maril, A., A. D. Wagner, and D. L. Schacter. 2001. "On the Tip of the Tongue: An Event-Related fMRI Study of Semantic Retrieval Failure and Cognitive Conflict." *Neuron* 31 (4): 653–60.

Markowitsch, H. J. 2003. "Psychogenic Amnesia." *Neuroimage* 20:S132–S138.

Marr, David. 1971. "Simple Memory: A Theory for Archicortex." *Philosophical Transactions of the Royal Society of London* 262 (841): 23–81. https://doi.org/10.1098/rstb.1971.0078.

Mason, Malia F., Michael I. Norton, John D. Van Horn, Daniel M. Wegner,

Scott T. Grafton, and C. Neil Macrae. 2007. "Wandering Minds: The Default Network and Stimulus-Independent Thought." *Science* 315 (5810): 393–95.

Maswood, Raeya, Christian C. Luhmann, and Suparna Rajaram. 2022. "Persistence of False Memories and Emergence of Collective False Memory: Collaborative Recall of DRM Word Lists." *Memory* 30 (4): 465–79.

Maswood, Raeya, and Suparna Rajaram. 2019. "Social Transmission of False Memory in Small Groups and Large Networks." *Topics in Cognitive Science* 11 (4): 687–709.

Mather, Mara. 2007. "Emotional Arousal and Memory Binding: An Object-Based Framework." *Perspectives on Psychological Science* 2 (1): 33–52.

Mather, Mara, and Laura L. Carstensen. 2005. "Aging and Motivated Cognition: The Positivity Effect in Attention and Memory." *Trends in Cognitive Sciences* 9 (10): 496–502.

Mather, Mara, David Clewett, Michiko Sakaki, and Carolyn W. Harley. 2016. "Norepinephrine Ignites Local Hotspots of Neuronal Excitation: How Arousal Amplifies Selectivity in Perception and Memory." *Behavioral and Brain Sciences* 39:e200.

Mayes, A., D. Montaldi, and E. Migo. 2007. "Associative Memory and the Medial Temporal Lobes." *Trends in Cognitive Sciences* 11 (3): 126–35.

McAdams, D. P. 2008. "Personal Narratives and the Life Story." In *Handbook of Personality: Theory and Research*, edited by O. P. John, R. W. Robins, and L. A. Pervin, 242–62. New York: Guilford Press.

McClelland, J. L., B. L. McNaughton, and R. C. O'Reilly. 1995. "Why There Are Complementary Learning Systems in the Hippocampus and Neocortex: Insights from the Successes and Failures of Connectionist Models of Learning and Memory." *Psychological Review* 102 (3): 419.

McClelland, J. L., D. E. Rumelhart, and PDP Research Group. 1986. *Parallel Distributed Processing*, vol. 2. Cambridge, MA: MIT Press.

McCloskey, Michael, and Neal J. Cohen. 1989. "Catastrophic Interference in Connectionist Networks: The Sequential Learning Problem." *Psychology of Learning and Motivation* 24:109–65.

McCulloch, Warren S., and Walter Pitts. 1943. "A Logical Calculus of the Ideas Immanent in Nervous Activity." *Bulletin of Mathematical Biophysics* 5 (4): 115–33.

McCullough, Andrew M., Maureen Ritchey, Charan Ranganath, and Andrew Yonelinas. 2015. "Differential Effects of Stress-Induced Cortisol Responses on Recollection and Familiarity-Based Recognition Memory." *Neurobiology of Learning and Memory* 123:1–10.

McDougle, Samuel D., Richard B. Ivry, and Jordan A. Taylor. 2016. "Taking Aim at the Cognitive Side of Learning in Sensorimotor Adaptation Tasks." *Trends in Cognitive Sciences* 20, no. 7: 535–44.

McEwen, Bruce S. 2007. "Physiology and Neurobiology of Stress and Adaptation: Central Role of the Brain." *Physiological Reviews* 87 (3): 873–904.

McEwen, Bruce S., Jay M. Weiss, and Leslie S. Schwartz. 1968. "Selective Retention of Corticosterone by Limbic Structures in Rat Brain." *Nature* 220 (5170): 911–12.

McGaugh, James L. 2018. "Emotional Arousal Regulation of Memory Consolidation." *Current Opinion in Behavioral Sciences* 19:55–60.

McIntyre, Christa K., James L. McGaugh, and Cedric L. Williams. 2012. "Interacting Brain Systems Modulate Memory Consolidation." *Neuroscience & Biobehavioral Reviews* 36 (7): 1750–62.

McKone, Elinor, Lulu Wan, Madeleine Pidcock, Kate Crookes, Katherine Reynolds, Amy Dawel, Evan Kidd, and Chiara Fiorentini. 2019. "A Critical Period for Faces: Other-Race Face Recognition Is Improved by Childhood but Not Adult Social Contact." *Scientific Reports* 9 (1): 1–13.

Meade, M. L., T. J. Nokes, and D. G. Morrow. 2009. "Expertise Promotes Facilitation on a Collaborative Memory Task." *Memory* 17 (1): 39–48.

Meade, Michelle L., and Henry L. Roediger. 2002. "Explorations in the Social Contagion of Memory." *Memory & Cognition* 30 (7): 995–1009.

Mednick, Sara. 2020. *The Power of the Downstate: Recharge Your Life Using Your Body's Own Restorative Systems.* New York: Hachette.

Mednick, S., and M. Ehrman. 2006. *Take a Nap! Change Your Life.* New York: Workman.

Mednick, S., K. Nakayama, and R. Stickgold. 2003. "Sleep-Dependent Learning: A Nap Is as Good as a Night." *Nature Neuroscience* 6 (7): 697–98.

Meissner, C. A., and J. C. Brigham. 2001. "Thirty Years of Investigating the Own-Race Bias in Memory for Faces: A Meta-analytic Review." *Psychology, Public Policy, and Law* 7 (1): 3.

Meister, M. L., and E. A. Buffalo. 2016. "Getting Directions from the Hippocampus: The Neural Connection Between Looking and Memory." *Neurobiology of Learning and Memory* 134:135–44.

Milivojevic, B., M. Varadinov, A. V. Grabovetsky, S. H. Collin, and C. F. Doeller. 2016. "Coding of Event Nodes and Narrative Context in the Hippocampus." *Journal of Neuroscience* 36 (49): 12412–24.

Milivojevic, B., A. Vicente-Grabovetsky, and C. F. Doeller. 2015. "Insight Reconfigures Hippocampal-Prefrontal Memories." *Current Biology* 25 (7): 821–30.

Miller, George A. 1956. "The Magic Number Seven Plus or Minus Two: Some Limits on Our Capacity for Processing Information." *Psychological Review* 63:9197.

Miller, Greg. 2007. "A Surprising Connection Between Memory and Imagination." *Science* 315 (5810): 312.

Miller, Jonathan F., Markus Neufang, Alec Solway, Armin Brandt, Michael Trippel, Irina Mader, Stefan Hefft, et al. (2013). "Neural Activity in Human Hippocampal Formation Reveals the Spatial Context of Retrieved Memories." *Science* 342 (6162): 1111–14.

Mineka, Susan, and John F. Kihlstrom. 1978. "Unpredictable and Uncontrollable Events: A New Perspective on Experimental Neurosis." *Journal of Abnormal Psychology* 87 (2): 256.

Minsky, Marvin. 1975. "A Framework for Representing Knowledge." MIT-AI Laboratory Memo 306, June 1974. Reprinted in P. Winston, ed., *The Psychology of Computer Vision*. New York: McGraw-Hill, 1975.

Mishkin, M., W. A. Suzuki, D. G. Gadian, and F. Vargha-Khadem. 1997. "Hierarchical Organization of Cognitive Memory." *Philosophical Transactions of the Royal Society of London. Series B: Biological Sciences* 352 (1360): 1461–67.

Montague, J. 2019. *Lost and Found: Memory, Identity, and Who We Become When We're No Longer Ourselves.* London: Hodder & Stoughton.

Montaldi, D., T. J. Spencer, N. Roberts, and A. R. Mayes. 2006. "The Neural System That Mediates Familiarity Memory." *Hippocampus* 16 (5): 504–20.

Mooney, Lindsey N., Elliott G. Johnson, Janani Prabhakar, and Simona Ghetti. 2021. "Memory-Related Hippocampal Activation during Sleep and Temporal Memory in Toddlers." *Developmental Cognitive Neuroscience* 47: 100908.

Moore, Christopher D., Michael X. Cohen, and Charan Ranganath. 2006. "Neural Mechanisms of Expert Skills in Visual Working Memory." *Journal of Neuroscience* 26 (43): 11187–96.

Moscovitch, Morris. 1989. "Confabulation and the Frontal Systems: Strategic Versus Associative Retrieval in Neuropsychological Theories of Memory." In *Varieties of Memory and Consciousness: Essays in Honour of Endel Tulving*, edited by H. L. Roediger and F. I. M. Craik, 133–60. Hillsdale, NJ: Lawrence Erlbaum.

Moscovitch, Morris, Roberto Cabeza, Gordon Winocur, and Lynn Nadel. 2016. "Episodic Memory and Beyond: The Hippocampus and Neocortex in Transformation." *Annual Review of Psychology* 67 (1): 105–34.

Moscovitch, Morris, and Gordon Winocur. 1992. "The Neuropsychology of Memory and Aging." *Handbook of Aging and Cognition* 315:372.

Mullan, Sean, and Wilder Penfield. 1959. "Illusions of Comparative Interpretation and Emotion: Production by Epileptic Discharge and by Electrical Stimulation in the Temporal Cortex." *AMA Archives of Neurology & Psychiatry* 81 (3): 269–84.

Münsterberg, Hugo. 1923. *On the Witness Stand: Essays on Psychology and Crime.* Clark Boardman.

Murayama, K., M. Matsumoto, K. Izuma, and K. Matsumoto. 2010. "Neural Basis of the Undermining Effect of Monetary Reward on Intrinsic Motivation." *Proceedings of the National Academy of Sciences* 107 (49): 20911–16.

Murayama, K., T. Miyatsu, D. Buchli, and B. C. Storm. 2014. "Forgetting as a Consequence of Retrieval: A Meta-analytic Review of Retrieval-Induced Forgetting." *Psychological Bulletin* 140 (5): 1383.

Murphy, C., V. Dehmelt, A. P. Yonelinas, C. Ranganath, and M. J. Gruber. 2021. "Temporal Proximity to the Elicitation of Curiosity Is Key for Enhancing Memory for Incidental Information." *Learning & Memory* 28 (2): 34–39.

Murphy, Gillian, Laura Lynch, Elizabeth Loftus, and Rebecca Egan. 2021. "Push Polls Increase False Memories for Fake News Stories." *Memory* 29 (6): 693–707.

Murray, E. A., S. P. Wise, and K. S. Graham. 2017. *The Evolution of Memory Systems: Ancestors, Anatomy, and Adaptations.* Oxford: Oxford University Press.

Nader, K., G. E. Schafe, and J. E. LeDoux. 2000. "Fear Memories Require Protein Synthesis in the Amygdala for Reconsolidation After Retrieval." *Nature* 406 (6797): 722–26.

Nadim, Farzan, and Dirk Bucher. 2014. "Neuromodulation of Neurons and Synapses." *Current Opinion in Neurobiology* 29:48–56.

Nattrass, Stuart, Darren P. Croft, Samuel Ellis, Michael A. Cant, Michael N. Weiss, Brianna M. Wright, Eva Stredulinsky, et al. 2019. "Postreproductive Killer Whale Grandmothers Improve the Survival of Their Grandoffspring." *Proceedings of the National Academy of Sciences* 116 (52): 26669–73.

Nauta, Walle J. H. 1971. "The Problem of the Frontal Lobe: A Reinterpretation." *Journal of Psychiatric Research* 8 (3): 167–87.

Navarrete, M., M. Valderrama, and P. A. Lewis. 2020. "The Role of Slow-Wave Sleep Rhythms in the Cortical-Hippocampal Loop for Memory Consolidation." *Current Opinion in Behavioral Sciences* 32:102–10.

Nelson, Katherine, and Robyn Fivush. 2004. "The Emergence of Autobiographical Memory: A Social Cultural Developmental Theory." *Psychological Review* 111 (2): 486.

Newell, Allen, John Calman Shaw, and Herbert A. Simon. 1958. "Chess-Playing Programs and the Problem of Complexity." *IBM Journal of Research and Development* 2 (4): 320–35.

Newman, David B., and Matthew E. Sachs. 2020. "The Negative Interactive Effects of Nostalgia and Loneliness on Affect in Daily Life." *Frontiers in Psychology* 11:2185.

Nielson, D. M., T. A. Smith, V. Sreekumar, S. Dennis, and P. B. Sederberg. 2015. "Human Hippocampus Represents Space and Time During Retrieval of Real-World Memories." *Proceedings of the National Academy of Sciences* 112 (35): 11078–83.

Nisbett, Richard E., and Timothy D. Wilson. 1977. "Telling More Than We Can Know: Verbal Reports on Mental Processes." *Psychological Review* 84 (3): 231–59.

Nobre, A. C., and G. McCarthy. 1995. "Language-Related Field Potentials in the Anterior-Medial Temporal Lobe: II. Effects of Word Type and Semantic Priming." *Journal of Neuroscience* 15 (2): 1090–98.

Nordahl, Christine Wu, Charan Ranganath, Andrew P. Yonelinas, Charles DeCarli, Evan Fletcher, and William J. Jagust. 2006. "White Matter Changes Compromise Prefrontal Cortex Function in Healthy Elderly Individuals." *Journal of Cognitive Neuroscience* 18 (3): 418–29.

Nordahl, Christine Wu, Charan Ranganath, Andrew P. Yonelinas, Charles DeCarli, Bruce R. Reed, and William J. Jagust. 2005. "Different Mechanisms of Episodic Memory Failure in Mild Cognitive Impairment." *Neuropsychologia* 43 (11): 1688–97.

Norman, Donald A., and Tim Shallice. 1986. "Attention to Action." In *Consciousness and Self-Regulation*, edited by Richard J. Davidson, Gary E. Schwartz, and David Shapiro, 1–18. Boston: Springer.

Norman, Kenneth A. 2010. "How Hippocampus and Cortex Contribute to Recognition Memory: Revisiting the Complementary Learning Systems Model." *Hippocampus* 20 (11): 1217–27.

Norman, K. A., E. L. Newman, and G. Detre. 2007. "A Neural Network Model of Retrieval-Induced Forgetting." *Psychological Review* 114 (4): 887.

Nyberg, L., R. Habib, A. R. McIntosh, and E. Tulving. 2000. "Reactivation of Encoding-Related Brain Activity During Memory Retrieval." *Proceedings of the National Academy of Sciences* 97 (20): 11120–24.

Oeberst, Aileen, Merle Madita Wachendörfer, Roland Imhoff, and Hartmut Blank. 2021. "Rich False Memories of Autobiographical Events Can Be Reversed." *Proceedings of the National Academy of Sciences* 118 (13).

O'Keefe, John, and Lynn Nadel. 1978. *The Hippocampus as a Cognitive Map.* Oxford: Oxford University Press.

———. 1979. "Précis of O'Keefe & Nadel's *The Hippocampus as a Cognitive Map.*" *Behavioral and Brain Sciences* 2 (4): 487–94.

Oliva, M. T., and B. C. Storm. 2022. "Examining the Effect Size and Duration of Retrieval-Induced Facilitation." *Psychological Research*, 1–14.

O'Reilly, R. C., Y. Munakata, M. J. Frank, T. E. Hazy, and Contributors. 2012. *Computational Cognitive Neuroscience.* Wiki Book, 4th ed., 2020. https://CompCogNeuro.org.

O'Reilly, Randall C., and Kenneth A. Norman. 2002. "Hippocampal and Neocortical Contributions to Memory: Advances in the Complementary Learning Systems Framework." *Trends in Cognitive Sciences* 6 (12): 505–10.

Owens, Justine, Gordon H. Bower, and John B. Black. 1979. "The 'Soap Opera' Effect in Story Recall." *Memory & Cognition* 7 (3): 185–91.

Paller, Ken A., Jessica D. Creery, and Eitan Schechtman. 2021. "Memory and

Sleep: How Sleep Cognition Can Change the Waking Mind for the Better." *Annual Review of Psychology* 72:123–50. https://doi.org/10.1146/annurev-psych-010419-050815.

Paller, Ken A., Gregory McCarthy, Elizabeth Roessler, Truett Allison, and Charles C. Wood. 1992. "Potentials Evoked in Human and Monkey Medial Temporal Lobe During Auditory and Visual Oddball Paradigms." *Electroencephalography and Clinical Neurophysiology/Evoked Potentials Section* 84 (3): 269–79.

Paller, Ken A., Joel L. Voss, and Stephen G. Boehm. 2007. "Validating Neural Correlates of Familiarity." *Trends in Cognitive Sciences* 11 (6): 243–50.

Palmqvist, Sebastian, Michael Schöll, Olof Strandberg, Niklas Mattsson, Erik Stomrud, Henrik Zetterberg, Kaj Blennow, Susan Landau, William Jagust, and Oskar Hansson. 2017. "Earliest Accumulation of β-amyloid Occurs Within the Default-Mode Network and Concurrently Affects Brain Connectivity." *Nature Communications* 8 (1): 1–13.

Palombo, Daniela J., Claude Alain, Hedvig Söderlund, Wayne Khuu, and Brian Levine. 2015. "Severely Deficient Autobiographical Memory (SDAM) in Healthy Adults: A New Mnemonic Syndrome." *Neuropsychologia* 72:105–18.

Pashler, H., N. Cepeda, R. V. Lindsey, E. Vul, and M. C. Mozer. 2009. "Predicting the Optimal Spacing of Study: A Multiscale Context Model of Memory." In *Advances in Neural Information Processing Systems* 22, edited by Y. Bengio, D. Schuurmans, J. Lafferty, C. K. I. Williams, and A. Culotta, 1321–29. La Jolla, CA: NIPS Foundation.

Pavlov, Ivan P. 1897. *The Work of the Digestive Glands.* London: Griffin.

———. 1924. "Lectures on the Work of the Cerebral Hemisphere, Lecture One." In Ivan Petrovich Pavlov, *Experimental Psychology and Other Essays.* New York: Philosophical Library, 1957.

———. 1927. *Conditioned Reflexes: An Investigation of the Physiological Activity of the Cerebral Cortex.* Oxford: Oxford University Press.

Peker, Müjde, and Ali I. Tekcan. 2009. "The Role of Familiarity Among Group Members in Collaborative Inhibition and Social Contagion." *Social Psychology* 40 (3): 111–18.

Pendergrast, M. 1996. *Victims of Memory: Sex Abuse Accusations and Shattered Lives.* Hinesburg, VT: Upper Access.

Penfield, Wilder. 1958. "Some Mechanisms of Consciousness Discovered During Electrical Stimulation of the Brain." *Proceedings of the National Academy of Sciences* 44 (2): 51–66.

Penfield, Wilder, and Brenda Milner. 1958. "Memory Deficit Produced by Bilateral Lesions in the Hippocampal Zone." *AMA Archives of Neurology & Psychiatry* 79 (5): 475–97.

Pennartz, C. M. A., R. Ito, P. F. M. J. Verschure, F. P. Battaglia, and T. W. Robbins. 2011. "The Hippocampal-Striatal Axis in Learning, Prediction and Goal-Directed Behavior." *Trends in Neurosciences* 34 (10): 548–59.

Pennycook, G., J. A. Cheyne, N. Barr, D. J. Koehler, and J. A. Fugelsang. 2015. "On the Reception and Detection of Pseudo-Profound Bullshit." *Judgment and Decision Making* 10 (6): 549–63.

Pennycook, G., and D. G. Rand. 2020. "Who Falls for Fake News? The Roles of Bullshit Receptivity, Overclaiming, Familiarity, and Analytic Thinking." *Journal of Personality* 88 (2): 185–200.

———. 2021. "The Psychology of Fake News." *Trends in Cognitive Sciences* 25 (5): 388–402.

Perry, C. J., I. Zbukvic, J. H. Kim, and A. J. Lawrence. 2014. "Role of Cues and Contexts on Drug-Seeking Behaviour." *British Journal of Pharmacology* 171 (20): 4636–72.

Peterson, Carole. 2002. "Children's Long-Term Memory for Autobiographical Events." *Developmental Review* 22 (3): 370–402.

Phelps, Elizabeth A. 2004. "Human Emotion and Memory: Interactions of the Amygdala and Hippocampal Complex." *Current Opinion in Neurobiology* 14 (2): 198–202.

Piaget, J. 1952. *The Origins of Intelligence in Children*. Translated by M. Cook. New York: W. W. Norton. https://doi.org/10.1037/11494-000.

Pichert, James W., and Richard C. Anderson. 1977. "Taking Different Perspectives on a Story." *Journal of Educational Psychology* 69 (4): 309.

Polich, John. 2007. "Updating P300: An Integrative Theory of P3a and P3b." *Clinical Neurophysiology* 118 (10): 2128–48.

Polyn, Sean M., Vaidehi S. Natu, Jonathan D. Cohen, and Kenneth A. Norman. 2005. "Category-Specific Cortical Activity Precedes Retrieval During Memory Search." *Science* 310 (5756): 1963–66.

Potts, R., G. Davies, and D. R. Shanks. 2019. "The Benefit of Generating Errors During Learning: What Is the Locus of the Effect?" *Journal of Experimental Psychology: Learning, Memory, and Cognition* 45 (6): 1023.

Prabhakar, Janani, Elliott G. Johnson, Christine Wu Nordahl, and Simona Ghetti. 2018. "Memory-Related Hippocampal Activation in the Sleeping Toddler." *Proceedings of the National Academy of Sciences* 115 (25): 6500–505.

Pribram, Karl H. 1973. "The Primate Frontal Cortex—Executive of the Brain." Chap. 14 in *Psychophysiology of the Frontal Lobes*, edited by K. H. Pribram and A. R. Luria, 293–314. New York: Academic Press.

Radvansky, Gabriel A., Abigail C. Doolen, Kyle A. Pettijohn, and Maureen Ritchey. 2022. "A New Look at Memory Retention and Forgetting." *Journal of Experimental Psychology: Learning, Memory, and Cognition* 48 (11): 1698–723.

Radvansky, Gabriel A., and Jeffrey M. Zacks. 2017. "Event Boundaries in Memory and Cognition." *Current Opinion in Behavioral Sciences* 17:133–40.

Raichle, Marcus E., Ann Mary MacLeod, Abraham Z. Snyder, William J. Powers, Debra A. Gusnard, and Gordon L. Shulman. 2001. "A Default Mode of Brain Function." *Proceedings of the National Academy of Sciences* 98 (2): 676–82.

Rajaram, Suparna. 2024. "Collaborative Remembering and Collective Memory." Chap. 75 in *The Oxford Handbook of Human Memory*, vol. 2, edited by M. J. Kahana and A. D. Wagner, 2169–92. Oxford: Oxford University Press.

Ranganath, Charan. 2010. "A Unified Framework for the Functional Organization of the Medial Temporal Lobes and the Phenomenology of Episodic Memory." *Hippocampus* 20 (11): 1263–90.

Ranganath, Charan, and Robert S. Blumenfeld. 2005. "Doubts About Double Dissociations Between Short- and Long-Term Memory." *Trends in Cognitive Sciences* 9 (8): 374–80.

Ranganath, Charan, Joe DeGutis, and Mark D'Esposito. 2004. "Category-Specific Modulation of Inferior Temporal Activity During Working Memory Encoding and Maintenance." *Cognitive Brain Research* 20 (1): 37–45.

Ranganath, Charan, and Mark D'Esposito. 2005. "Directing the Mind's Eye: Prefrontal, Inferior and Medial Temporal Mechanisms for Visual Working Memory." *Current Opinion in Neurobiology* 15 (2): 175–82.

Ranganath, Charan, and Liang-Tien Hsieh. 2016. "The Hippocampus: A Special Place for Time." *Annals of the New York Academy of Sciences* 1369 (1): 93–110.

Ranganath, Charan, Marcia K. Johnson, and Mark D'Esposito. 2000. "Left Anterior Prefrontal Activation Increases with Demands to Recall Specific Perceptual Information." *Journal of Neuroscience* 20 (22): RC108.

———. 2003. "Prefrontal Activity Associated with Working Memory and Episodic Long-Term Memory." *Neuropsychologia* 41 (3): 378–89.

Ranganath, Charan, and Ken A. Paller. 1999. "Frontal Brain Potentials During Recognition Are Modulated by Requirements to Retrieve Perceptual Detail." *Neuron* 22 (3): 605–13.

———. 2000. "Neural Correlates of Memory Retrieval and Evaluation." *Cognitive Brain Research* 9 (2): 209–22.

Ranganath, Charan, and Gregor Rainer. 2003. "Neural Mechanisms for Detecting and Remembering Novel Events." *Nature Reviews Neuroscience* 4 (3): 193–202.

Ranganath, Charan, and Maureen Ritchey. 2012. "Two Cortical Systems for Memory-Guided Behavior." *Nature Reviews Neuroscience* 13 (10): 713–26.

Ranganath, Charan, Andrew P. Yonelinas, Michael X. Cohen, Christine J. Dy, Sabrina M. Tom, and Mark D'Esposito. 2004. "Dissociable Correlates of

Recollection and Familiarity Within the Medial Temporal Lobes." *Neuropsychologia* 42 (1): 2–13.

Ranganath, Charan. 2024. "Episodic Memory." Chap. 6 in *The Oxford Handbook of Human Memory*, vol. 1, edited by M. J. Kahana and A. D. Wagner, 151–71. Oxford: Oxford University Press.

Rasch, B., C. Büchel, S. Gais, and J. Born. 2007. "Odor Cues During Slow-Wave Sleep Prompt Declarative Memory Consolidation." *Science* 315 (5817): 1426–29.

Rauers, A., M. Riediger, F. Schmiedek, and U. Lindenberger. 2011. "With a Little Help from My Spouse: Does Spousal Collaboration Compensate for the Effects of Cognitive Aging?" *Gerontology* 57 (2): 161–66.

Reagh, Zachariah M., Angelique I. Delarazan, Alexander Garber, and Charan Ranganath. 2020. "Aging Alters Neural Activity at Event Boundaries in the Hippocampus and Posterior Medial Network." *Nature Communications* 11 (1): 1–12.

Reagh, Zachariah M., and Charan Ranganath. 2023. "Flexible Reuse of Cortico-hippocampal Representations During Encoding and Recall of Naturalistic Events." *Nature Communications* 14 (1): 1279.

Rendell, L., and H. Whitehead. 2001. "Culture in Whales and Dolphins." *Behavioral and Brain Sciences* 24 (2): 309–24.

Rescorla, R. A., and R. I. Solomon. 1967. "Two-Process Learning Theory: Relationships Between Pavlovian Conditioning and Instrumental Learning." *Psychological Review* 74 (3): 151–82. https://doi.org/10.1037/h0024475.

Riccio, D. C., P. M. Millin, and A. R. Bogart. 2006. "Reconsolidation: A Brief History, a Retrieval View, and Some Recent Issues." *Learning & Memory* 13 (5): 536–44.

Richland, L. E., N. Kornell, and L. S. Kao. 2009. "The Pretesting Effect: Do Unsuccessful Retrieval Attempts Enhance Learning?" *Journal of Experimental Psychology: Applied* 15 (3): 243.

Ritchey, Maureen, Andrew M. McCullough, Charan Ranganath, and Andrew P. Yonelinas. 2017. "Stress as a Mnemonic Filter: Interactions Between Medial Temporal Lobe Encoding Processes and Post-Encoding Stress." *Hippocampus* 27 (1): 77–88.

Ritchey, Maureen, Maria E. Montchal, Andrew P. Yonelinas, and Charan Ranganath. 2015. "Delay-Dependent Contributions of Medial Temporal Lobe Regions to Episodic Memory Retrieval." *eLife* 4:e05025.

Ritchey, Maureen, Shao-Fang Wang, Andrew P. Yonelinas, and Charan Ranganath. 2019. "Dissociable Medial Temporal Pathways for Encoding Emotional Item and Context Information." *Neuropsychologia* 124:66–78.

Robbins, T. W., and B. J. Everitt. 2007. "A Role for Mesencephalic Dopamine in Activation: Commentary on Berridge (2006)." *Psychopharmacology* 191 (3): 433–37.

Roediger, Henry L., III. 1985. "Remembering Ebbinghaus." Review of H. Ebbinghaus, *Memory: A Contribution to Experimental Psychology*. *Contemporary Psychology* 30 (7): 519–23. https://doi.org/10.1037/023895.

———. 1990. "Implicit Memory: Retention Without Remembering." *American Psychologist* 45 (9): 1043.

———. 2003. "Bartlett, Frederic Charles." In *Encyclopedia of Cognitive Science*, edited by Lynn Nadel, 1:319–22. Hoboken, NJ: Wiley.

———. 2008. "Relativity of Remembering: Why the Laws of Memory Vanished." *Annual Review of Psychology* 59:225–54.

Roediger, Henry L., III., and Andrew C. Butler. 2011. "The Critical Role of Retrieval Practice in Long-Term Retention." *Trends in Cognitive Sciences* 15 (1): 20–27.

Roediger, Henry L., III, and Jeffrey D. Karpicke. 2006. "Test-Enhanced Learning: Taking Memory Tests Improves Long-Term Retention." *Psychological Science* 17 (3): 249–55.

Roediger, Henry L., III, and Kathleen B. McDermott. 1995. "Creating False Memories: Remembering Words Not Presented in Lists." *Journal of Experimental Psychology: Learning, Memory, and Cognition* 21 (4): 803.

———. 1996. "False Perceptions of False Memories." *Journal of Experimental Psychology: Learning, Memory, and Cognition* 22 (3): 814–16. https://doi.org/10.1037/0278-7393.22.3.814.

Roediger, Henry L., III, Michelle L. Meade, and Erik T. Bergman. 2001. "Social Contagion of Memory." *Psychonomic Bulletin & Review* 8 (2): 365–71.

Rowland, Christopher A. 2014. "The Effect of Testing Versus Restudy on Retention: A Meta-analytic Review of the Testing Effect." *Psychological Bulletin* 140 (6): 1432.

Rubin, R. D., S. Brown-Schmidt, M. C. Duff, D. Tranel, and N. J. Cohen. 2011. "How Do I Remember That I Know You Know That I Know?" *Psychological Science* 22 (12) (December): 1574–82. https://doi.org/10.1177/0956797611418245.

Rudoy, J. D., J. L. Voss, C. E. Westerberg, and K. A. Paller. 2009. "Strengthening Individual Memories by Reactivating Them During Sleep." *Science* 326 (5956): 1079.

Rugg, M. D., and K. L. Vilberg. 2013. "Brain Networks Underlying Episodic Memory Retrieval." *Current Opinion in Neurobiology* 23 (2): 255–60.

Rugg, M. D., and E. L. Wilding. 2000. "Retrieval Processing and Episodic Memory." *Trends in Cognitive Sciences* 4 (3): 108–15.

Rumelhart, David E., and Andrew Ortony. 1977. "The Representation of Knowledge in Memory." In *Schooling and the Acquisition of Knowledge*, edited by R. C. Anderson, R. J. Spiro, and W. E. Montague, 99–135. Hillsdale, NJ: Lawrence Erlbaum.

Ryan, Jennifer D., Robert R. Althoff, Stephen Whitlow, and Neal J. Cohen. 2000. "Amnesia Is a Deficit in Relational Memory." *Psychological Science* 11 (6): 454–61.

Ryan, Jennifer D., and Kelly Shen. 2020. "The Eyes Are a Window into Memory." *Current Opinion in Behavioral Sciences* 32:1–6.

Saletan, W. 2000. "Push Me, Poll You." *Slate,* February 15, 2000. https://slate.com/news-and-politics/2000/02/push-me-poll-you.html.

Sanders, K. E., S. Osburn, K. A. Paller, and M. Beeman. 2019. "Targeted Memory Reactivation During Sleep Improves Next-Day Problem Solving." *Psychological Science* 30 (11): 1616–24.

Sapolsky, Robert M. 1994. *Why Zebras Don't Get Ulcers: A Guide to Stress, Stress-Related Diseases, and Coping.* New York: W. H. Freeman.

———. 2002. "Endocrinology of the Stress-Response." In *Behavioral Endocrinology,* edited by J. B. Becker, S. M. Breedlove, D. Crews, and M. M. McCarthy, 409–50. Cambridge, MA: MIT Press.

———. 2003. "Taming Stress." *Scientific American* 289 (3): 86–95.

Sazma, M. A., G. S. Shields, and A. P. Yonelinas. 2019. "The Effects of Post-Encoding Stress and Glucocorticoids on Episodic Memory in Humans and Rodents." *Brain and Cognition* 133:12–23.

Schacter, Daniel L. 2002. *The Seven Sins of Memory: How the Mind Forgets and Remembers.* Boston: Mariner Books.

———. 2022. "Media, Technology, and the Sins of Memory." *Memory, Mind & Media* 1:e1.

Schacter, Daniel L., and Donna Rose Addis. 2007. "The Cognitive Neuroscience of Constructive Memory: Remembering the Past and Imagining the Future." *Philosophical Transactions of the Royal Society B: Biological Sciences* 362 (1481): 773–86.

Schacter, Daniel L., Donna Rose Addis, and Randy L. Buckner. 2008. "Episodic Simulation of Future Events: Concepts, Data, and Applications." *Annals of the New York Academy of Sciences* 1124 (1): 39–60.

Schacter, Daniel L., R. L. Buckner, W. Koutstaal, A. M. Dale, and B. R. Rosen. 1997. "Late Onset of Anterior Prefrontal Activity During True and False Recognition: An Event-Related fMRI Study." *Neuroimage* 6 (4): 259–69.

Schacter, Daniel L., Mieke Verfaellie, and Dan Pradere. 1996. "The Neuropsychology of Memory Illusions: False Recall and Recognition in Amnesic Patients." *Journal of Memory and Language* 35 (2): 319–34.

Schank, R. C., and R. P. Abelson. 1977. *Scripts, Plans, Goals and Understanding: An Inquiry into Human Knowledge Structures.* Hillsdale, NJ: Lawrence Erlbaum.

Schiller, D., and E. A. Phelps. 2011. "Does Reconsolidation Occur in Humans?" *Frontiers in Behavioral Neuroscience* 5:24.

Schlag, A. K. 2020. "Percentages of Problem Drug Use and Their Implications for Policy Making: A Review of the Literature." *Drug Science, Policy and Law* 6. https://doi.org/10.1177/2050324520904540.

Schulkind, Matthew D., Laura Kate Hennis, and David C. Rubin. 1999. "Music, Emotion, and Autobiographical Memory: They're Playing Your Song." *Memory & Cognition* 27 (6): 948–55.

Schultz, Wolfram. 1997. "Dopamine Neurons and Their Role in Reward Mechanisms." *Current Opinion in Neurobiology* 7 (2): 191–97.

———. 2006. "Behavioral Theories and the Neurophysiology of Reward." *Annual Review of Psychology* 87:115.

Scoboria, Alan, Giuliana Mazzoni, Irving Kirsch, and Leonard S. Milling. 2002. "Immediate and Persisting Effects of Misleading Questions and Hypnosis on Memory Reports." *Journal of Experimental Psychology: Applied* 8 (1): 26.

Scoboria, Alan, Kimberley A. Wade, D. Stephen Lindsay, Tanjeem Azad, Deryn Strange, James Ost, and Ira E. Hyman. 2017. "A Mega-analysis of Memory Reports from Eight Peer-Reviewed False Memory Implantation Studies." *Memory* 25 (2): 146–63.

Scoville, William Beecher, and Brenda Milner. 1957. "Loss of Recent Memory After Bilateral Hippocampal Lesions." *Journal of Neurology, Neurosurgery, and Psychiatry* 20 (1): 11.

Shaw, Julia, and Stephen Porter. 2015. "Constructing Rich False Memories of Committing Crime." *Psychological Science* 26 (3): 291–301.

Sheldon, Signy, Can Fenerci, and Lauri Gurguryan. 2019. "A Neurocognitive Perspective on the Forms and Functions of Autobiographical Memory Retrieval." *Frontiers in Systems Neuroscience* 13:4.

Shields, Grant S., Andrew M. McCullough, Maureen Ritchey, Charan Ranganath, and Andrew P. Yonelinas. 2019. "Stress and the Medial Temporal Lobe at Rest: Functional Connectivity Is Associated with Both Memory and Cortisol." *Psychoneuroendocrinology* 106:138–46.

Shields, Grant S., Matthew A. Sazma, Andrew M. McCullough, and Andrew P. Yonelinas. 2017. "The Effects of Acute Stress on Episodic Memory: A Meta-analysis and Integrative Review." *Psychological Bulletin* 143 (6): 636.

Shields, Grant S., Matthew A. Sazma, and Andrew P. Yonelinas. 2016. "The Effects of Acute Stress on Core Executive Functions: A Meta-analysis and Comparison with Cortisol." *Neuroscience & Biobehavioral Reviews* 68:651–68.

Silvia, Paul J. 2008. "Interest—the Curious Emotion." *Current Directions in Psychological Science* 17 (1): 57–60.

Silvia, Paul J., and Alexander P. Christensen. 2020. "Looking Up at the Curious Personality: Individual Differences in Curiosity and Openness to Experience." *Current Opinion in Behavioral Sciences* 35:1–6.

Simon, H. A. 1974. "How Big Is a Chunk? By Combining Data from Several Experiments, a Basic Human Memory Unit Can Be Identified and Measured." *Science* 183 (4124): 482–88.

Simons, Jon S., Jane R. Garrison, and Marcia K. Johnson. 2017. "Brain Mechanisms of Reality Monitoring." *Trends in Cognitive Sciences* 21 (6): 462–73.

Sinclair, A. H., and M. D. Barense. 2019. "Prediction Error and Memory Reactivation: How Incomplete Reminders Drive Reconsolidation." *Trends in Neurosciences* 42 (10): 727–39.

Singh, D., K. A. Norman, and A. C. Schapiro. 2022. "A Model of Autonomous Interactions Between Hippocampus and Neocortex Driving Sleep-Dependent Memory Consolidation." *Proceedings of the National Academy of Sciences* 119 (44): e2123432119.

Smallwood, Jonathan, and Jonathan W. Schooler. 2015. "The Science of Mind Wandering: Empirically Navigating the Stream of Consciousness." *Annual Review of Psychology* 66:487–518.

Soares, Julia S., and Benjamin C. Storm. 2018. "Forget in a Flash: A Further Investigation of the Photo-Taking-Impairment Effect." *Journal of Applied Research in Memory and Cognition* 7 (1): 154–60.

Sokolov, E. N. 1963. "Higher Nervous Functions: The Orienting Reflex." *Annual Review of Physiology* 25 (1): 545–80.

———. 1990. "The Orienting Response, and Future Directions of Its Development." *Pavlovian Journal of Biological Science* 25 (3): 142–50.

Solomon, M., A.-M. Iosif, M. K. Krug, C. W. Nordahl, E. Adler, C. Mirandola, and S. Ghetti. 2019. "Emotional False Memory in Autism Spectrum Disorder: More Than Spared." *Journal of Abnormal Psychology* 128 (4): 352–63. https://doi.org/10.1037/abn0000418.

Soltani, M., and R. T. Knight. 2000. "Neural Origins of the P300." *Critical Reviews in Neurobiology* 14 (3–4).

Sporns, Olaf. 2010. *Networks of the Brain*. Cambridge, MA: MIT Press.

Squire, Larry R. 1986. "Mechanisms of Memory." *Science* 232 (4758): 1612–19.

Squire, Larry R., and S. M. Zola. 1998. "Episodic Memory, Semantic Memory, and Amnesia." *Hippocampus* 8 (3): 205–11.

Squires, Nancy K., Kenneth C. Squires, and Steven A. Hillyard. 1975. "Two Varieties of Long-Latency Positive Waves Evoked by Unpredictable Auditory Stimuli in Man." *Electroencephalography and Clinical Neurophysiology* 38 (4): 387–401.

Staniloiu, Angelica, and Hans J. Markowitsch. 2014. "Dissociative Amnesia." *Lancet Psychiatry* 1 (3): 226–41.

Stapleton, J. M., and E. Halgren. 1987. "Endogenous Potentials Evoked in Simple Cognitive Tasks: Depth Components and Task Correlates." *Electroencephalography and Clinical Neurophysiology* 67 (1): 44–52.

Stare, Christopher J., Matthias J. Gruber, Lynn Nadel, Charan Ranganath, and Rebecca L. Gómez. 2018. "Curiosity-Driven Memory Enhancement Persists over Time but Does Not Benefit from Post-learning Sleep." *Cognitive Neuroscience* 9 (3–4): 100–15.

Staresina, B. P., T. O. Bergman, M. Bonnefond, R. van der Meij, O. Jensen, L. Deuker, C. E. Elger, N. Axmacher, and J. Fell. 2015. "Hierarchical Nesting of Slow Oscillations, Spindles and Ripples in the Human Hippocampus During Sleep." *Nature Neuroscience* 18 (11): 1679–86.

Stawarczyk, David, Christopher N. Wahlheim, Joset A. Etzel, Abraham Z. Snyder, and Jeffrey M. Zacks. 2020. "Aging and the Encoding of Changes in Events: The Role of Neural Activity Pattern Reinstatement." *Proceedings of the National Academy of Sciences* 117 (47): 29346–53.

Stemerding, L. E., D. Stibbe, V. A. van Ast, and M. Kindt. 2022. "Demarcating the Boundary Conditions of Memory Reconsolidation: An Unsuccessful Replication." *Scientific Reports* 12 (1): 2285.

Stern, Chantal E., Suzanne Corkin, R. Gilberto González, Alexander R. Guimaraes, John R. Baker, Peggy J. Jennings, Cindy A. Carr, Robert M. Sugiura, Vasanth Vedantham, and Bruce R. Rosen. 1996. "The Hippocampal Formation Participates in Novel Picture Encoding: Evidence from Functional Magnetic Resonance Imaging." *Proceedings of the National Academy of Sciences* 93, no. 16: 8660–65.

Stiffler, L. 2011. "Understanding Orca Culture." *Smithsonian Magazine*, August 2011. https://www.smithsonianmag.com/science-nature/understanding-orca-culture-12494696/.

St. Jacques, Peggy L. 2012. "Functional Neuroimaging of Autobiographical Memory." Chap. 7 in *Understanding Autobiographical Memory*, edited by Dorthe Berntsen and David C. Rubin, 114–38. Cambridge: Cambridge University Press.

Stuss, D. T., F. I. Craik, L. Sayer, D. Franchi, and M. P. Alexander. 1996. "Comparison of Older People and Patients with Frontal Lesions: Evidence from Word List Learning." *Psychology and Aging* 11 (3): 387.

Sutton, S., M. Braren, J. Zubin, and E. R. John. 1965. "Evoked Potential Correlates of Stimulus Uncertainty." *Science* 150:1187–88.

Swallow, Khena M., Deanna M. Barch, Denise Head, Corey J. Maley, Derek Holder, and Jeffrey M. Zacks. 2011. "Changes in Events Alter How People Remember Recent Information." *Journal of Cognitive Neuroscience* 23, no. 5: 1052–64.

Swallow, K. M., J. M. Zacks, and R. A. Abrams. 2009. "Event Boundaries in Perception Affect Memory Encoding and Updating." *Journal of Experimental Psychology: General* 138 (2): 236.

Szpunar, K. K., J. M. Watson, and K. B. McDermott. 2007. "Neural Substrates of Envisioning the Future." *Proceedings of the National Academy of Sciences* 104 (2): 642–47.

Takeuchi, Tomonori, Adrian J. Duszkiewicz, and Richard G. M. Morris. 2014. "The Synaptic Plasticity and Memory Hypothesis: Encoding, Storage and Persistence." *Philosophical Transactions of the Royal Society B: Biological Sciences* 369 (1633): 20130288.

Tambini, A., N. Ketz, and L. Davachi. 2010. "Enhanced Brain Correlations During Rest Are Related to Memory for Recent Experiences." *Neuron* 65 (2): 280–90.

Teuber, H. L. 1964. "The Riddle of Frontal Lobe Function in Man." In *The Frontal Granular Cortex and Behavior*, edited by J. M. Warren and K. Akert. New York: McGraw-Hill.

Teyler, T. J., and P. DiScenna. 1986. "The Hippocampal Memory Indexing Theory." *Behavioral Neuroscience* 100 (2): 147.

Teyler, T. J., and J. W. Rudy. 2007. "The Hippocampal Indexing Theory and Episodic Memory: Updating the Index." *Hippocampus* 17 (12): 1158–69.

Thakral, Preston P., Aleea L. Devitt, Nadia M. Brashier, and Daniel L. Schacter. 2021. "Linking Creativity and False Memory: Common Consequences of a Flexible Memory System." *Cognition* 217:104905.

Thakral, Preston P., Kevin P. Madore, Sarah E. Kalinowski, and Daniel L. Schacter. 2020. "Modulation of Hippocampal Brain Networks Produces Changes in Episodic Simulation and Divergent Thinking." *Proceedings of the National Academy of Sciences* 117 (23): 12729–40.

Thomas, Ayanna K., John B. Bulevich, and Elizabeth F. Loftus. 2003. "Exploring the Role of Repetition and Sensory Elaboration in the Imagination Inflation Effect." *Memory & Cognition* 31 (4): 630–40.

Thomas, Ayanna K., and Elizabeth F. Loftus. 2002. "Creating Bizarre False Memories Through Imagination." *Memory & Cognition* 30 (3): 423–31.

Tolman, Edward C. 1948. "Cognitive Maps in Rats and Men." *Psychological Review* 55 (4): 189.

Tulving, Endel. 1972. "Episodic and Semantic Memory." Chap. 12 in *Organization of Memory*, edited by E. Tulving and W. Donaldson, 381–403. New York: Academic Press.

———. 1985. "Memory and Consciousness." *Canadian Psychology / Psychologie canadienne* 26 (1): 1–12.

Tulving, Endel, Hans J. Markowitsch, Shitij Kapur, Reza Habib, and Sylvain Houle. 1994. "Novelty Encoding Networks in the Human Brain: Positron Emission Tomography Data." *NeuroReport: International Journal for Rapid Communication of Research in Neuroscience* 5, no. 18: 2525–28.

Tulving, Endel, and Daniel L. Schacter. 1990. "Priming and Human Memory Systems." *Science* 247 (4940): 301–6.

Tupes, E. C., and R. E. Christal. 1992. "Recurrent Personality Factors Based on Trait Ratings." *Journal of Personality* 60 (2): 225–51.

Umbach, G., P. Kantak, J. Jacobs, M. Kahana, B. E. Pfeiffer, M. Sperling, and B. Lega. 2020. "Time Cells in the Human Hippocampus and Entorhinal Cortex Support Episodic Memory." *Proceedings of the National Academy of Sciences* 117 (45): 28463–74.

Uncapher, Melina R., and Anthony D. Wagner. 2018. "Minds and Brains of Media Multitaskers: Current Findings and Future Directions." *Proceedings of the National Academy of Sciences* 115 (40): 9889–96.

Unkelbach, Christian, Alex Koch, Rita R. Silva, and Teresa Garcia-Marques. 2019. "Truth by Repetition: Explanations and Implications." *Current Directions in Psychological Science* 28 (3): 247–53.

Vargha-Khadem, F., and F. Cacucci. 2021. "A Brief History of Developmental Amnesia." *Neuropsychologia* 150:107689.

Vargha-Khadem, F., D. G. Gadian, K. E. Watkins, A. Connelly, W. Van Paesschen, and M. Mishkin. 1997. "Differential Effects of Early Hippocampal Pathology on Episodic and Semantic Memory." *Science* 277 (5324): 376–80.

Vinogradova, Olga S. 2001. "Hippocampus as Comparator: Role of the Two Input and Two Output Systems of the Hippocampus in Selection and Registration of Information." *Hippocampus* 11 (5): 578–98.

Viscogliosi, C., H. Asselin, S. Basile, K. Borwick, Y. Couturier, M. J. Drolet, D. Gagnon, et al. 2020. "Importance of Indigenous Elders' Contributions to Individual and Community Wellness: Results from a Scoping Review on Social Participation and Intergenerational Solidarity." *Canadian Journal of Public Health* 111 (5): 667–81.

Volkow, Nora D., Joanna S. Fowler, Gene-Jack Wang, James M. Swanson, and Frank Telang. 2007. "Dopamine in Drug Abuse and Addiction: Results of Imaging Studies and Treatment Implications." *Archives of Neurology* 64 (11): 1575–79.

Voss, J. L., D. J. Bridge, N. J. Cohen, and J. A. Walker. 2017. "A Closer Look at the Hippocampus and Memory." *Trends in Cognitive Sciences* 21 (8): 577–88.

Voss, Michelle W., Carmen Vivar, Arthur F. Kramer, and Henriette van Praag. 2013. "Bridging Animal and Human Models of Exercise-Induced Brain Plasticity." *Trends in Cognitive Sciences* 17 (10): 525–44.

Wade, Kimberley A., Maryanne Garry, and Kathy Pezdek. 2018. "Deconstructing Rich False Memories of Committing Crime: Commentary on Shaw and Porter (2015)." *Psychological Science* 29 (3): 471–76.

Wagner, Anthony D. 1999. "Working Memory Contributions to Human Learning and Remembering." *Neuron* 22 (1): 19–22.

Walker, Matthew. 2017. *Why We Sleep: Unlocking the Power of Sleep and Dreams.* New York: Simon & Schuster.

Wamsley, E. J., and R. Stickgold. 2011. "Memory, Sleep, and Dreaming: Experiencing Consolidation." *Sleep Medicine Clinics* 6 (1): 97–108.

Wang, M. Z., and B. Y. Hayden. 2019. "Monkeys Are Curious About Counterfactual Outcomes." *Cognition* 189:1–10.

Wang, Shao-Fang, Maureen Ritchey, Laura A. Libby, and Charan Ranganath. 2016. "Functional Connectivity Based Parcellation of the Human Medial Temporal Lobe." *Neurobiology of Learning and Memory* 134:123–34.

Wang, Wei-Chun, Michele M. Lazzara, Charan Ranganath, Robert T. Knight, and Andrew P. Yonelinas. 2010. "The Medial Temporal Lobe Supports Conceptual Implicit Memory." *Neuron* 68 (5): 835–42.

Wang, Wei-Chun, Charan Ranganath, and Andrew P. Yonelinas. 2014. "Activity Reductions in Perirhinal Cortex Predict Conceptual Priming and Familiarity-Based Recognition." *Neuropsychologia* 52:19–26.

Watson, John B. 1913. "Psychology as the Behaviorist Views It." *Psychological Review* 20 (2): 158.

Weldon, M. S., and K. D. Bellinger. 1997. "Collective Memory: Collaborative and Individual Processes in Remembering." *Journal of Experimental Psychology: Learning, Memory, and Cognition* 23 (5): 1160.

West, Robert L. 1996. "An Application of Prefrontal Cortex Function Theory to Cognitive Aging." *Psychological Bulletin* 120 (2): 272.

Wheeler, Mark E., Steven E. Petersen, and Randy L. Buckner. 2000. "Memory's Echo: Vivid Remembering Reactivates Sensory-Specific Cortex." *Proceedings of the National Academy of Sciences* 97 (20): 11125–29.

Wilding, E. L., and M. D. Rugg. 1996. "An Event-Related Potential Study of Recognition Memory with and Without Retrieval of Source." *Brain* 119 (3): 889–905.

Wilson, Robert C., Amitai Shenhav, Mark Straccia, and Jonathan D. Cohen. 2019. "The Eighty-Five Percent Rule for Optimal Learning." *Nature Communications* 10 (1): 1–9.

Wise, Roy A. 2004. "Dopamine, Learning and Motivation." *Nature Reviews Neuroscience* 5 (6): 483–94.

Wise, Roy A., and Mykel A. Robble. 2020. "Dopamine and Addiction." *Annual Review of Psychology* 71:79–106.

Wittmann, Bianca C., Nico Bunzeck, Raymond J. Dolan, and Emrah Düzel. 2007. "Anticipation of Novelty Recruits Reward System and Hippocampus While Promoting Recollection." *Neuroimage* 38 (1): 194–202.

Wixted, J. T. 2007. "Dual-Process Theory and Signal-Detection Theory of Recognition Memory." *Psychological Review* 114 (1): 152.

Wolf, Oliver T. 2009. "Stress and Memory in Humans: Twelve Years of Progress?" *Brain Research* 1293:142–54.

Wright, L. A., L. Horstmann, E. A. Holmes, and J. I. Bisson. 2021. "Consolidation/Reconsolidation Therapies for the Prevention and Treatment of PTSD and Re-experiencing: A Systematic Review and Meta-Analysis." *Translational Psychiatry* 11 (1): 453.

Xia, Chenjie. 2006. "Understanding the Human Brain: A Lifetime of Dedicated Pursuit. Interview with Dr. Brenda Milner." *McGill Journal of Medicine* 9 (2): 165.

Xie, Lulu, Hongyi Kang, Qiwu Xu, Michael J. Chen, Yonghong Liao, Meenakshisundaram Thiyagarajan, John O'Donnell, et al. 2013. "Sleep Drives Metabolite Clearance from the Adult Brain." *Science* 342 (6156): 373–77.

Xue, Gui, Qi Dong, Chuansheng Chen, Zhonglin Lu, Jeanette A. Mumford, and Russell A. Poldrack. 2010. "Greater Neural Pattern Similarity Across Repetitions Is Associated with Better Memory." *Science* 330 (6000): 97–101.

Yonelinas, Andrew P. 1994. "Receiver-Operating Characteristics in Recognition Memory: Evidence for a Dual-Process Model." *Journal of Experimental Psychology: Learning, Memory, and Cognition* 20 (6): 1341.

———. 2001. "Consciousness, Control, and Confidence: The 3 Cs of Recognition Memory." *Journal of Experimental Psychology: General* 130 (3): 361–79.

———. 2002. "The Nature of Recollection and Familiarity: A Review of 30 Years of Research." *Journal of Memory and Language* 46 (3): 441–517.

Yonelinas, A. P., M. Aly, W.-C. Wang, and J. D. Koen. 2010. "Recollection and Familiarity: Examining Controversial Assumptions and New Directions." *Hippocampus* 20 (11): 1178–94.

Yonelinas, A. P., N. E. Kroll, J. R. Quamme, M. M. Lazzara, M. J. Sauvé, K. F. Widaman, and R. T. Knight. 2002. "Effects of Extensive Temporal Lobe Damage or Mild Hypoxia on Recollection and Familiarity." *Nature Neuroscience* 5 (11): 1236–41.

Yonelinas, Andrew P., and Collen M. Parks. 2007. "Receiver Operating Characteristics (ROCs) in Recognition Memory: A Review." *Psychological Bulletin* 133 (5): 800.

Yonelinas, Andrew P., Colleen M. Parks, Joshua D. Koen, Julie Jorgenson, and Sally P. Mendoza. 2011. "The Effects of Post-encoding Stress on Recognition Memory: Examining the Impact of Skydiving in Young Men and Women." *Stress* 14 (2): 136–44.

Yonelinas, A. P., C. Ranganath, A. D. Ekstrom, and B. J. Wiltgen. 2019. "A Contextual Binding Theory of Episodic Memory: Systems Consolidation Reconsidered." *Nature Reviews Neuroscience* 20 (6): 364–75.

Yonelinas, Andrew P., and Maureen Ritchey. 2015. "The Slow Forgetting of Emotional Episodic Memories: An Emotional Binding Account." *Trends in Cognitive Sciences* 19 (5): 259–67.

Yoo, H. B., G. Umbach, and B. Lega. 2021. "Neurons in the Human Medial Temporal Lobe Track Multiple Temporal Contexts During Episodic Memory Processing." *NeuroImage* 245:118689.

Zacks, Jeffrey M. 2020. "Event Perception and Memory." *Annual Review of Psychology* 71:165–91.

Zacks, J. M., T. S. Braver, M. A. Sheridan, D. I. Donaldson, A. Z. Snyder, J. M. Ollinger, R. L. Buckner, and M. E. Raichle. 2001. "Human Brain Activity Time-Locked to Perceptual Event Boundaries." *Nature Neuroscience* 4 (6): 651–55.

Zacks, Jeffrey M., and Barbara J. Tversky. 2001. "Event Structure in Perception and Conception." *Psychological Bulletin* 127 (1): 3.

Zacks, Rose T., and Lynn Hasher. 2006. "Aging and Long-Term Memory: Deficits Are Not Inevitable." In *Lifespan Cognition: Mechanisms of Change*, edited by E. Bialystok and F. I. M. Craik, 162–77. Oxford: Oxford University Press.

Zadra, A., and R. Stickgold. 2021. *When Brains Dream*. New York: W. W. Norton.

Zajonc, R. B. 1968. "Attitudinal Effects of Mere Exposure." *Journal of Personality and Social Psychology* 9 (2, pt. 2): 1–27.

———. 2001. "Mere Exposure: A Gateway to the Subliminal." *Current Directions in Psychological Science* 10 (6): 224–28.

Zheng, Yicong, Xiaonan L. Liu, Satoru Nishiyama, Charan Ranganath, and Randall C. O'Reilly. 2022. "Correcting the Hebbian Mistake: Toward a Fully Error-Driven Hippocampus." *PLoS Computational Biology* 18 (10): e1010589.

INDEX